AP CHEMISTRY

Frederick J. Rowe
Northport High School
Northport, NY

With the assistance of:
Sherry Berman-Robinson
Carl Sandburg High School
Orland Park, IL

Prentice Hall
New York • London • Toronto • Sydney • Tokyo • Singapore

DEDICATED TO the more than 25,000 students and the AP chemistry teachers who toil yearly to master the college level chemistry course which is called AP Chemistry. Take the challenge of a foreign language, add a good dose of required mathematical skill, and top it all off with many abstract concepts and you have the Advanced Placement Chemistry syllabus.

Second Edition

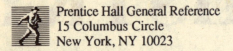

Prentice Hall General Reference
15 Columbus Circle
New York, NY 10023

Copyright © 1993, 1990 by Frederick J. Rowe
All rights reserved
including the right of reproduction
in whole or in part in any form

An Arco Book

Arco, Prentice Hall, and colophons are
registered trademarks of Simon & Schuster, Inc.

Library of Congress Cataloging-in-Publication Data

Rowe, Frederick J.
    AP chemistry / Frederick J. Rowe; with the
assistance of Sherry Berman-Robinson.—2nd ed.
    p. cm.
At head of title: Arco.
"An Arco book"—T.p. verso.
Includes index.
ISBN 0-671-84778-3
1. Chemistry—Examinations, questions, etc. 2. College
entrance achievement tests. I. Berman-Robinson, Sherry.
II. Title. III. Title: Arco AP chemistry.
QD42.R7   1993   93-18320
540'.76—dc20   CIP

Manufactured in the United States of America

1 2 3 4 5 6 7 8 9 10

Contents

	Acknowledgments	viii
	Introduction	ix
Part I	**Review of Selected Topics**	
Chapter 1	**Atomic Structure and Periodicity**	
	1. Atomic Theory and Atomic Structure	1
	2. Quantum Numbers and Atomic Orbitals	2
	3. Determining Electron Structure	3
	4. Ion Structure	5
	5. Relationships in the Periodic Table	5
	6. Nuclear Chemistry	8
Chapter 2	**Chemical Bonding**	
	1. Chemical Bonding	9
	2. Octet Formulas and Lewis Dots	10
	3. Expect the Unexpected (Lewis Formulas and the Octet Rule)	12
	Multiple Lewis Structures of the Same Compound	13
	Resonance	13
	Odd-Electron Molecules	14
	Molecules with Incomplete Octets	14
	Central Atoms with More Than an Octet	14
	4. Shapes of Molecules and Ions (VSEPR Theory)	15
	5. Attractive Forces	18
	6. Molecular Orbitals	21
	7. Hybrid Orbitals, Sigma Bonding, and Pi Bonding	23
	8. Coordination Chemistry	24
Chapter 3	**Stoichiometry**	
	1. The Mole Concept and Molar Mass	27
	2. Empirical and Molecular Formulas	28
	3. Reaction Stoichiometry	30
	4. Limiting Reactants	31
	5. Molarity, Molality, and Mole Fraction	32
	6. Solution Stoichiometry	34
	7. Colligative Properties	34

Chapter 4   **States of Matter**

1.	The Solid, Liquid, and Gas States	36
2.	Changes of State	38
3.	Kinetic Molecular Theory of Gases; Graham's Law	39
4.	The Ideal Gas Law	40
5.	Dalton's Law of Partial Pressures	42
6.	Properties of Solutions; Raoult's Law	42

Chapter 5   **Reaction Kinetics**

1.	Rate Equations	44
2.	First Order Reactions	48
3.	Activation Energy and Catalysis	49
4.	Heat Effects and the Arrhenius Equation	51
5.	Reaction Mechanism	53

Chapter 6   **Equilibrium**

1.	Equilibrium Constants	55
2.	Concentration Changes in Equilibrium Reactions	56
3.	LeChatelier's Principle	58
4.	Temperature Change and the van't Hoff Equation	59
5.	Solubility Equilibria and K_{sp}	61
6.	Common Ion Effect	62
7.	Selective Precipitation	63
8.	Mixtures of Two Solutions and 'Bounce Back'	64
9.	Strong Acids and Strong Bases	66
10.	Weak Acids, Weak Bases, and pH	68
11.	Buffers	71
12.	Titration and Neutralization	72
13.	The Titration of a Weak Acid with a Strong Base	74

Chapter 7 **Thermodynamics**

    1. Calorimetry, Internal Energy, and Enthalpy — 76
    2. Hess' Law and Heat of Reaction — 78
    3. Heats of Formation, Combustion, and Reaction — 80
    4. Bond Energy — 81
    5. Entropy — 82
    6. Free Energy — 83
    7. Free Energy and Equilibrium — 84
    8. Free Energy Change and Net Cell Potential — 86

Chapter 8 **Electrochemistry**

    1. Balancing Oxidation–Reduction Reactions — 87
    2. The Faraday and Electrolytic Cells — 90
    3. Voltaic (Chemical) Cells — 94
    4. Standard Reduction Potentials — 96
    5. The Nernst Equation — 98

Part II Multiple-Choice Questions

About Section I: The Multiple-Choice Questions — 101

Set 1 Atomic Structure and Periodicity — 102
Answer Key and Explanations — 106

Set 2 Chemical Bonding — 108
Answer Key and Explanations — 111

Set 3 Stoichiometry — 113
Answer Key and Explanations — 117

Set 4 States of Matter — 120
Answer Key and Explanations — 124

Set 5 Reaction Kinetics — 126
Answer Key and Explanations — 131

Set 6 Equilibrium — 133
Answer Key and Explanations — 137

Set 7	**Thermodynamics**	141
	Answer Key and Explanations	146
Set 8	**Electrochemistry**	149
	Answer Key and Explanations	153

Part III — Free Response Section

About Section II: Free Response Problems — 159

Chapter 1. Part A Sample Problems

1. Gaseous Equilibrium — 161
2. Solubility Equilibrium — 161
3. Acid-Base Equilibrium — 162
4. Acid-Base Titration — 162

Sample Answers and Explanations

1. Gaseous Equilibrium — 163
2. Solubility Equilibrium — 164
3. Acid-Base Equilibrium — 165
4. Acid-Base Titration — 166

Chapter 2. Part B Sample Problems

1. Electrochemistry—Electrolytic Cells — 168
2. Electrochemistry—Voltaic Cells — 168
3. Kinetics — 169
4. Stoichiometry—Empirical Formulas and Colligative Properties — 169
5. Stoichiometry—Standard Solutions — 170
6. Thermodynamics — 170

Sample Answers and Explanations

1. Electrochemistry—Electrolytic Cells — 171
2. Electrochemistry—Voltaic Cells — 171
3. Kinetics — 172
4. Stoichiometry—Empirical Formulas and Colligative Properties — 172
5. Stoichiometry—Standard Solutions — 173
6. Thermodynamics — 174

Chapter 3. Part C: Equations

1.	General Considerations	175
2.	Solubility Rules for Salts in Water Solutions	176
3.	Use of the Solubility Rules-Precipitation Reactions	177
4.	Reactions Involving No Changes in Oxidation States	178
	Replacement Reactions	178
	Metathesis Reactions (Acid and Base Anhydrides)	178
	Decomposition Reactions	179
	Hydrolysis Reactions	179
	Reactions of Coordination Compounds and Ions	180
	Nonaqueous Definitions of Acids and Bases	180
5.	Redox Reactions—Oxidation State Changes by the Oxidizing and the Reducing Agents	181
	Redox Reactions Between an Oxidizing and a Reducing Agent	182
	Redox Displacement Reactions	182
	Redox Combination Reactions	183
	Redox Decomposition Reactions	183
6.	Practice Equation Writing	183
	Sample Answers and Explanations	185

Chapter 4. Part D: Essays — 189

Sample Essay Questions — 191
Sample Answers and Explanations — 196

Part IV Appendix

A.	Summary of Formulas	203
B.	Standard Reduction Potentials	206
C.	Vapor Pressure of Water	207
D.	Constants	207
	Index	208
	Periodic Table of the Elements	212

Acknowledgments

The sales success of the first edition of this book proves that an AP Chemistry review book has been needed for some time. Publishers had been justifiably reluctant to take on such a project since the mechanicals for chemistry books are expensive. It requires the extensive use of subscripts, superscripts, diagrams, equations and charts to clearly present the information.

This review book was written, edited, designed, illustrated, and typeset using a Macintosh® computer using desktop publishing techniques. The printing mechanicals were produced using a laser printer.

I again wish to thank Linda Bernbach, Senior Editor of ARCO books, for recognizing the potential of this project from the initial proposal. She has enthusiastically guided the idea from approval by the publisher toward producing the second edition of a book that is as accurate and attractive as possible.

Sherry Berman-Robinson of Carl Sandburg High School (Orland Park, IL) has a special talent for successfully preparing students for the AP examination. She and her students make a special effort to prepare for the Part C Equations. Sherry has shared some of her secrets by writing the first draft of Chapter 3 in Part III.

Bob Sims of the Westminster School (Atlanta, GA) has extensive expertise as a chemistry teacher and as the chair of the ACS Examinations Institute High School Advanced Chemistry test committee (1988 and 1990). He has contributed to this effort with a careful critique of the review chapters.

Helen Stone, formerly of Ben L. Smith High School (Greensboro, NC), also has served as the chair of the ACS Examinations Institute High School Advanced Chemistry test committee (1986 and 1992). She spent many hours carefully reviewing the questions and wore out many red pens in this effort. Her many suggestions are incorporated into the multiple choice questions, problems, and essays.

A special thanks to Doris—my wife, friend and colleague —who shares the experience of many years coping with the needs of Advanced Placement students (AP Studio in Art at Northport High School). Her support and philosophy have been indispensable, as always.

Introduction

This second edition of this book is written in response to a need to have in one place materials that students can use to efficiently prepare for the examination. Each page of the second edition was carefully reviewed and edited. In response to student requests, additional topics are reviewed in Part I and additional questions have been added to Part III.

A careful investigation of the concepts that should be mastered is the focus of the review. A distillation of current college chemistry courses cannot replace the more complete coverage of topics that is possible during a year's study using a 700-page textbook.

It is neither required, nor expected, that candidates will have mastered the concepts covered in all of the texts. Students awarded the highest score on the AP test do not answer every question correctly. In fact, candidates who answer a little more than half of the Section I multiple choice questions correctly—depending on an equivalent success with the Section II free-response questions—can anticipate their effort will be rewarded with a score of 5.0.

The book consists of three parts. The examples, questions, and problems are modeled in the AP Chemistry style, but they are not questions which have been used in the examination.

Part I: Review of Selected Topics

Part II: Multiple-Choice Questions:
Strategy, questions, answers and explanations

Part III: Free Response Section II:
Strategy, questions, answers and explanations

Practice questions in addition to those supplied in this book are easy to get. The College Board has published the:

"Set of free-response questions used in recent years"
(Item 254932...$2.00)

"Entire 1984 AP Chemistry Examination and Key"
(Item 468223...$5.00)

"Entire 1989 AP Chemistry Examination and Key"
(Item 468255...$10.00)

Write to the:

Advanced Placement Program
CN 6670
Princeton, NJ 08541-6670

The American Chemical Society prepares and publishes standardized multiple-choice tests. The proper level for AP preparation is the multiple choice tests published biennially by the ACS Examinations Institute entitled "High School Chemistry (Advanced)". The exams are confidential, so a teacher or school official must order and supervise the use of the more recent 1988 ADV, 1990 ADV or 1992 ADV examinations. Earlier examinations are not as confidential. The author has been selected to be the chairman of the 1994 ADV examination now in preparation.

Some students contend that the *High School Chemistry (Advanced)* questions are more rigorous than those of Section I of the AP examination.

Write or call the:

ACS DivCHED Examinations Institute
223 Brackett Hall
Clemson University
Clemson, SC 29634
(803) 656-1249
(803) 656-1250 (FAX)

The explanations in this book use problem-solving techniques that can be applied to questions dealing with several topics. Current examination usage of units is maintained. A unit like molarity, M, is written in the exam as "mol L^{-1}." This is read as "moles per Liter."

The AP examination is a race against the clock. It is assumed that the candidate who purchases this book will, like an athlete, make sure to have the right equipment. This includes extensive practice in learning how to use—and bringing to the exam—a hand-held, non-programmable scientific calculator. The calculator should be capable of—in addition to performing the usual addition, multiplication, etc.—computing exponential numbers, logarithms (both ln and log), powers and roots. No provision is made in the review to explain how to 'do-it-by-hand'.

Part I

Review of Selected Topics

Chapter 1
Atomic Structure and Periodicity

1. **Atomic Theory and Atomic Structure**

   An atom is electrically neutral when it contains the same number of protons and electrons. The number of protons (the atomic number) determines the identity of the element. The mass number is the sum of the number of protons and neutrons.

   ### Example

Element	Protons	Electrons	Neutrons	Mass #
In	49	49	71	120
Sn	50	50	70	120
Sb	51	51	70	121
Sb	51	51	69	120
Te	52	52	68	120

   Atoms of a particular element may vary in mass (isotopes) because of a varied number of neutrons in the nucleus. Particular isotopes are identified by the symbol of the element with the atomic number at the lower-left corner and the mass number at the upper left corner.

   ### Example

Element	Protons	Neutrons	Mass #	Symbol
Sn	50	69	119	$^{119}_{50}\text{Sn}$
Sn	50	70	120	$^{120}_{50}\text{Sn}$
Sb	51	70	121	$^{121}_{51}\text{Sb}$
Sb	51	69	120	$^{120}_{51}\text{Sb}$
Te	52	68	120	$^{120}_{52}\text{Te}$
Te	52	72	124	$^{124}_{52}\text{Te}$

The weighted-average of the naturally occurring isotopes of an element determines the chemical atomic weight.

Example

Element	Mass #	Abundance, %	Mass # × Abundance
Cu	62.9296	69.20	4355
Cu	64.927	30.80	2000
Total		100.00	6355

Atomic weight = Weighted Average = Total ÷ 100% = 6355 ÷ 100.00 = **63.55**

Atoms may gain or lose electrons to form electrically charged species called 'ions'.

Example

Element	Protons	Electrons	Mass #	Ionic Symbol
Na	11	10	23	$^{23}_{11}Na^{1+}$
Ga	31	28	70	$^{70}_{31}Ga^{3+}$
S	16	18	35	$^{35}_{16}S^{2-}$
Br	35	36	81	$^{81}_{35}Br^{1-}$
Pb	82	78	206	$^{206}_{82}Pb^{4+}$

2. Quantum Numbers and Atomic Orbitals

The energies of electrons in atoms are quantized, which means that electrons in atoms have only certain allowed energy states. The energy of an electron in an atom is determined mainly by the value of n, the principal quantum number. The principal quantum number defines a shell, at some average distance from the nucleus, capable of holding electrons having about the same energy. The value of the principal quantum number, n, can be any positive integer. The lowest energy electrons have a principal quantum number of 1 ($n=1$). Only values up to $n=7$ are necessary to describe the electrons in the first 118 elements.

The shells consist of one or more subshells determined by the orbital quantum number, l. Allowed values of l include any positive integer from zero (0) to $(n-1)$. Usually, instead of integers, values of l are assigned the letters s, p, d, and f. Orbital quantum number (l) values of 0, 1, 2, and 3 are all that are necessary to list the subshells of the known or soon to be known first 118 elements.

Not only does the orbital quantum number indicate differences in energy of electrons within a shell, but it also indicates the shape of the orbital.

Subshell, l	0	1	2	3
Orbital Designation	s	p	d	f
Shape	spherical	dumbbell	N.A.	N.A.

Note: N.A. means not appropriate for testing at the AP level.

Each subshell consists of one or more orbitals of the same energy indicated by the magnetic quantum number, m_l. Permitted values of m_l are integers from $-l$ to $+l$.

Example
The table gives a summary of orbitals for the first four periods.

Shell n	Subshell l	Orbital Designation	Magnetic QN m_l	No. of Orbitals
1	0	1s	0	1
2	0	2s	0	1
	1	2p	-1 0 +1	3
3	0	3s	0	1
	1	3p	-1 0 +1	3
	2	3d	-2 -1 0 +1 +2	5
4	0	4s	0	1
	1	4p	-1 0 +1	3
	2	4d	-2 -1 0 +1 +2	5
	3	4f	-3 -2 -1 0 +1 +2 +3	7

Electrons behave as if they were spinning. Two electrons can occupy the same orbital, but they must spin in opposite directions—the Pauli Exclusion Principle. The electron spin is indicated by the spin quantum number, m_s, which has values of $\pm\frac{1}{2}$.

3. **Determining Electron Structure**

The electronic configuration of any atom in its ground (most stable and lowest energy) state can be determined by using the 'aufbau' (building up) procedure. Two electrons are added per orbital (lowest energy orbitals first) until the number of added electrons equals the number of protons in the atom (making the atoms electrically neutral).

Electrons placed in a subshell consisting of more than one orbital will go into an empty orbital rather than pair up with the first electron in a half-filled orbital (Hund's Rule). And its spin direction will be the same as that of the first electron (parallel spin). Electron configurations and the order of filling subshells can readily be obtained from a periodic table.

Note: The principal quantum number, n, of an element's valence shell is the same as the horizontal row—the period—where the element is found.

Block	Groups	Configuration
s-block:	1A through 2A	ns^1 through ns^2
p-block:	3A through 7A plus Group 0 (Noble Gases)	$ns^2 np^1$ through $ns^2 np^6$
d-block:	Transition Elements 3B through 2B	***$(n-1)d^1 ns^2$ through ***$(n-1)d^{10} ns^2$
f-block:	Inner Transition Elements Lanthanide and Actinide Eeries	$(n-2)f^1 (n-1)d^1 ns^2$ through $(n-2)f^{14} (n-1)d^1 ns^2$

***Some irregularity exists, particularly in Groups 1B and 6B.

The periodic table shows how the s-, p-, d-, and f-blocks are positioned.

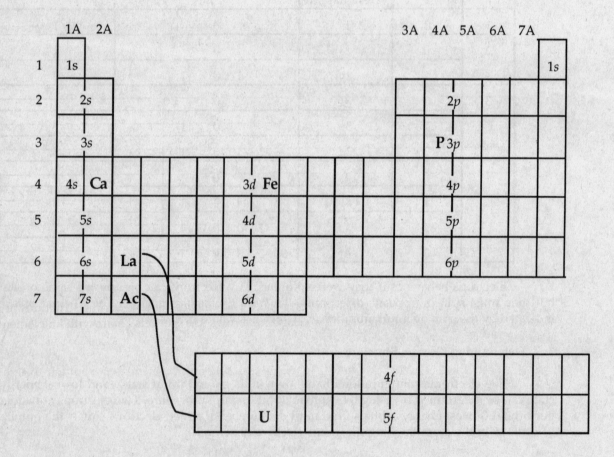

Example

Use the periodic table on page 4 to determine the outer electron structure for calcium (Ca), iron (Fe), phosphorus (P), and uranium (U).

Calcium is $[Ar]4s^2$. Calcium is found in the second box of the s-block of period 4.

Iron is $[Ar]3d^6\,4s^2$. Iron is located in the sixth box of the d-block of period 4, after passing through the $4s$-block. The outer, $4s$-electrons are written last, even though they are filled before the $3d$-electrons

Phosphorus is $[Ne]3s^2\,3p^3$. Phosphorus is found in the third box of the p-block of period 3, after passing through the $3s$-block.

Uranium is $[Rn]5f^3\,6d^1\,7s^2$. Uranium is located in the third box of the f-block of period 7, after passing through the $7s$-block and through the first box of the $6d$-block.

4. Ion Structure

Electron configurations for ions can be predicted from their respective atoms. Positive ions are formed when highest-energy (generally the outermost) electrons are lost. Negative ions are formed when electrons are added to the outermost subshell.

Example

Element	Structure	Ion	Structure
Ca	$[Ar]4s^2$	Ca^{2+}	$[Ar]$ or $[Ne]3s^23p^6$
S	$[Ne]3s^23p^4$	S^{2-}	$[Ne]3s^23p^6$ or $[Ar]$
Mn	$[Ar]3d^54s^2$	Mn^{2+}	$[Ar]3d^5$
Ti	$[Ar]3d^24s^2$	Ti^{3+}	$[Ar]3d^1$
Ag	$[Kr]4d^{10}5s^1$	Ag^+	$[Kr]4d^{10}$

5. Relationships in the Periodic Table

Elements having similar chemical properties and valence shell electronic structures are placed in vertical columns called 'groups'. Horizontal rows, the 'periods', start with an element belonging to Group 1A, the Alkali Metals.

The radii of atoms and ions decrease as the atomic number increases from left to right across a period. The cause is an increased attraction between the increased number of protons in the nucleus and electrons in the valence shell. There is a discontinuity at Group 5A, where a change from positive to negative ion formation occurs.

1. *Positive* ions are *smaller* than their respective atom.
2. *Negative* ions are *larger* than their respective atom.

The radii of atoms and ions increase as atomic numbers increase from top to bottom down a group. This is due to an increased number of shells and the shielding effect of all the inner electrons.

Ionization energy is a measure of the tightness with which electrons are held by an atom or ion. It is the amount of energy required to remove an electron from an isolated, neutral atom in the gaseous state. The ionization energy of atoms and ions increases as the atomic number increases from left to right across a period. There is a discontinuity where there is a change from positive to negative ion formation in the middle of the periodic table.

1. Electrons are held *more* tightly by positive ions.
2. Electrons are held *less* tightly by negative ions.

The ionization energy of atoms and ions decreases as the atomic number increases from the top to the bottom of a group.

Electron affinity is the energy released when an electron is added to an atom, generally a nonmetal. The electron affinity of atoms generally follows the same trend as the ionization energy.

1. Electrons are strongly held by an element with a *large* negative value of the electron affinity.
2. Nitrogen, N_2, and the Noble Gases are relatively non-reactive. These elements have little attraction for additional electrons and have *small* electron affinities (zero or positive values).
3. Oxygen, O, and fluorine, F, are small atoms with little room for additional electrons and have electron affinities which are *less* than expected.

The table shows the **electron affinities** of some nonmetals, in kilo Joules mole^{-1}.

IIIA	IVA	VA	VIA	VIIA	0
B	C	N	O	F	Ne
−23	−123	0	−141	−322	+29
	Si	P	S	Cl	Ar
	−120	−74	−200	−349	+35
	Ge	As	Se	Br	Kr
	−116	−77	−195	−324	+39
		Sb	Te	I	Xe
		−101	−190	−295	+40

This table shows the **periodic trends** of ionization energy, electron affinity, atomic radius, and metallic character.

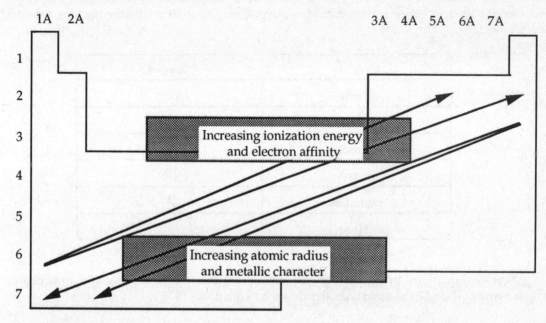

Example

Use the position of the elements in the periodic table to arrange each set of three elements in order of *increasing* size, ionization energy, and electron affinity.

a. Be, Mg, Ca
b. S, Cl, F
c. As, N, F
d. Ga, Ge, In
e. Te, I, Xe

Increasing Size	Increasing Ionization Energy	Increasing Electron Affinity
Be < Mg < Ca	Ca < Mg < Be	Ca < Mg < Be
F < Cl < S	S < Cl < F	S < F < Cl
F < N < As	As < N < F	N < As < F
Ge < Ga < In	In < Ga < Ge	In < Ga < Ge
Xe < I < Te	Te < I < Xe	Xe < Te < I

6. Nuclear Chemistry

When a natural radioactive atom decays, its nucleus loses either an alpha or a beta particle. If the nucleus of an atom is bombarded by an alpha particle, a beta particle, a neutron, a proton a deuteron, or a triton, the bombarding particle may be captured by the nucleus, forming a new element.

Particle	Symbol
alpha	^4_2He
beta	$^0_{-1}e$
neutron	^1_0n
proton	^1_1H
deuteron	^2_1H
triton	^3_1H

Nuclear equations are written by recognizing that the sum of the atomic numbers and the mass number must be the same on both sides of the equation.

Example

$$^{14}_7N + ^4_2He \rightarrow ^{18}_9F \rightarrow ^{17}_8O + ^1_1H$$

The rate of radioactive decay is expressed in terms of half-life. The same equations used for radioactive decay are used for a first order chemical reaction in chemical kinetics.

$$\ln \frac{N_t}{N_o} = -\lambda t \qquad \lambda = \frac{0.693}{t_{1/2}} \qquad \ln \frac{N_t}{N_o} = -\frac{0.693\,t}{t_{1/2}}$$

N_t is the number of moles of the isotope at time t.
N_o is the original number of moles of the isotope.
$\frac{N_t}{N_o}$ is the mole fraction of the isotope remaining at time t.
λ is the decay constant and $t_{1/2}$ is the half-life.

Example

Radioactive sodium-24 has a half-life of 15.0 hours. How many days will it take for the radioactivity to fall to 0.10 of its original activity?

$$\ln 0.10 = -\frac{0.693\,t}{15.0 \text{ hours}}$$

Answer: Time (t) = 50. hours = **2.1 days**

Chapter 2
Chemical Bonding

1. **Chemical Bonding**

Our current model of the chemical bond makes use of the definitions of the three strong bonds between atoms or ions: the metallic, the ionic, and the covalent.

Metallic Bonding

The bonding between metals must account for the typical physical properties associated with metals:

...uniform, high electrical and heat conductivity

...high malleability and ductility.

...shininess and luster.

The simpler 'electron-sea' definition is still useful, and describes the metallic bond as being cations or positive ions (the 'kernel') surrounded by a 'sea of mobile electrons'. Mobile electrons will explain the easy movement of heat and electricity. The 'kernel' can be easily realigned during compression or tension. The 'sea of mobile electrons' reflects light in many directions, causing metals to be shiny.

Ionic Bonding

The formation of bonds between ions must account for the typical physical properties associated with salts:

...brittleness.

...lack of electrical conductivity in the solid state.

...solubility in polar solvents such as water.

...conductivity in aqueous solutions.

When a metal reacts with a nonmetal, the metal loses its valence electrons and the nonmetal gains these electrons. The valence shell of the nonmetal is completed and gives it a stable noble gas electron structure. The cations and anions formed are held together by electrostatic attraction. A strongly bonded crystalline structure of ions explains the brittleness and the lack of electrical conductivity in the solid state. The existence of ions accounts for the attraction to polar solvents. Dissolved ions are mobile charges and can conduct electricity.

Covalent Bonding

Nonmetals have strong attractions for electrons as evidenced by their high ionization energy, electronegativity and electron affinity. A covalent bond is the sharing of a pair of electrons between two nonmetallic atoms. These electrons may be shared equally or unequally. Unequal sharing of electrons is predicted by an electronegativity difference and the result is a polar covalent bond with some ionic character.

2. Octet Formulas and Lewis Dots

There are over two-million known molecular compounds, so it makes sense to know how to visualize and represent covalently bonded substances. Many molecules form electron-pair bonds such that each atom achieves a closed-shell electron configuration. Hydrogen needs two electrons to achieve a closed-shell and acquire the stable, noble-gas structure of helium. The other nonmetals, such as fluorine, require eight electrons, *an octet*, to achieve electron stability.

Hydrogen: H• + •H ⇒ H:H

Fluorine: :F• + •F: ⇒ :F:F:

In writing the Lewis structure for a compound, include only the outermost (or valence) electrons. For the atoms of the representative group elements, the number of valence electrons is the same as the group number.

The covalent bond between atoms is ordinarily shown as a line between bonded atoms. The unshared electron-pairs (commonly called lone pair electrons) are shown as dots. The example shows four common compounds containing a period 2 nonmetal and hydrogen. The central atom in each case is surrounded by an octet of electrons, either in pairs of shared electrons (a line) or in lone pairs (two dots).

Example

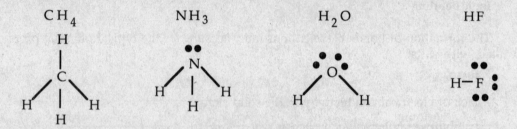

Lewis structures may be written by following the simple rules illustrated in the example.

Example

Draw a possible Lewis electron dot diagram for CH_4S.

1.	Draw a structure for the molecule by joining atoms by single bonds at first.	H | H—C—S—H | H
2.	Count the number of valence electrons. Add electrons for negative ions. Subtract electrons for positive ions.	C + 4H + 1S = ? 4 + 4 + 6 = **14** valence electrons
3.	Deduct two electrons for each bond.	14 − (5 × 2) = **4** unbonded electrons
4.	Distribute the remaining electrons as lone pairs to give each atom an 'octet' of electrons.	H | H—C—S̈—H | H
5.	If there are too few electrons for 'octets', consider multiple-bonding.	

Example: Structural formulas with single bonds

Formula	$POCl_3$	ClO_2^-
Step 1 Line structure	Cl | Cl—P—O | Cl	$[O-Cl-O]^-$
Step 2 Valence electrons	1P + 3Cl + 1O = ? 5 + 21 + 6 = **32**	1Cl + 2O + charge = ? 7 + 12 + 1 = **20**
Step 3 Unbonded electrons	32 − (4 × 2) = **24**	20 − (2 × 2) = **16**
Step 4 Lewis electron-dot diagram	:C̈l: | :C̈l—P—Ö: | :C̈l:	$[:\ddot{O}-\ddot{C}l-\ddot{O}:]^-$

Example: Structural formulas with multiple bonds

Formula	ClNO	CO$_3^{2-}$
Step 1 Line structure	Cl–N(–O) with N at top	[O–C(–O)(–O)]$^{2-}$
Step 2 Valence electrons	1Cl + 1N + 1O = ? 7 + 5 + 6 = 18	1C + 3O + charge = ? 4 + 18 + 2 = 24
Step 3 Unbonded electrons	18 − (2 × 2) = 14	24 − (3 × 2) = 18
Step 4 Lewis electron-dot diagram	Cl–N–O with lone pairs	[O–C(–O)(–O)]$^{2-}$ with lone pairs
The number of unbonded electrons is too small to allow an octet of electrons to be placed around each of the atoms. A double covalent bond solves the problem.		
Step 5 Unbonded electrons	18 − (3 × 2) = 12	24 − (4 × 2) = 16
Step 6 Line structure	Cl–N=O with lone pairs	[O–C(=O)(–O)]$^{2-}$ with lone pairs

Lewis structures can correctly show the difference in the bonding between an ionic compound and a covalent compound. In the electron-dot diagram of *ionic* compounds:
1. Positive ions are shown with *no* valence electrons.
2. Negative ions are shown in brackets with 8 valence electrons.
3. The hydride ion, H$^-$, is shown in brackets with 2 valence electrons.

Examples
The first of the two compounds in each pair of the example is the *covalent* compound.

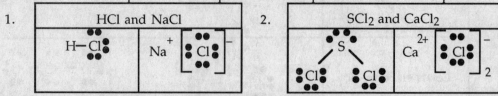

3. Expect the Unexpected (Lewis Formulas and the Octet Rule)

The rules for Lewis structures may be followed explicitly to allow the correct representation of most molecules and ions. A number of unexpected answers come up often enough to warrant further discussion. The 'answer' can be resolved only with additional information concerning the physical and chemical properties of the molecule being described.

Multiple Lewis Structures of the Same Compound

Diazomethane has the structural formula H₂CNN. The diagrams show three Lewis structures possible for the compound.

$$\text{H}_2\text{C}=\text{N}=\text{N} \quad\quad \text{H}-\overset{|}{\underset{H}{\text{C}}}-\text{N}\equiv\text{N} \quad\quad :\text{C}\equiv\text{N}-\overset{|}{\underset{H}{\text{N}}}-\text{H}$$

Resonance

When the rules for Lewis structures are applied to SO_2, SO_3, NO_2^- or NO_3^-, two or three equivalent structures result. The properties of these structures indicate that all the bonds are the same. The phenomenon of resonance occurs when more than one structure can be drawn, but where one structure is not enough to describe the species.

Example:

Sulfur dioxide, SO_2, and nitrate ion, NO_3^-

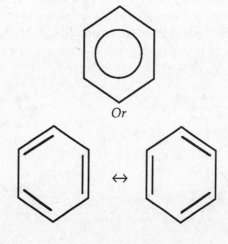

Benzene, C_6H_6

Odd-Electron Molecules

NO$_2$, CN, and NO are molecules with an odd total number of valence electrons. There will be at least one unpaired electron in the structure and at least one atom lacking an octet of electrons. These molecules with unpaired electrons will be 'paramagnetic', as contrasted with 'diamagnetic' molecules, whose electrons are all paired.

NO$_2$

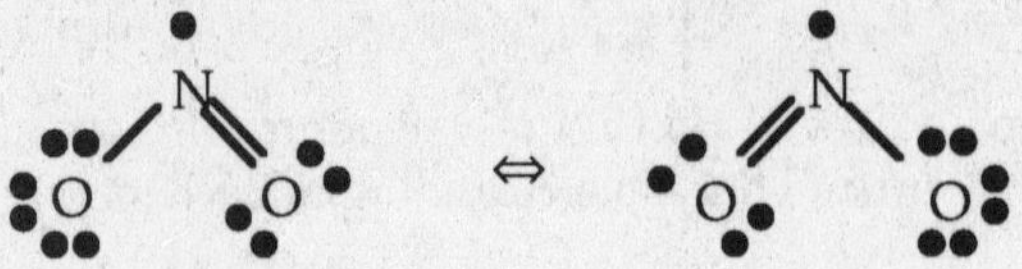

CN

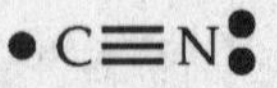

Molecules with Incomplete Octets

Beryllium and boron are representative of elements which can form covalent bonds, but whose central atom only has two or three valence electrons, respectively. BeCl$_2$ and BF$_3$ have structures with less than an octet of electrons.

BeCl$_2$

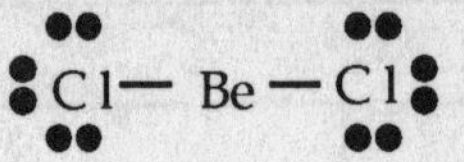

BF$_3$

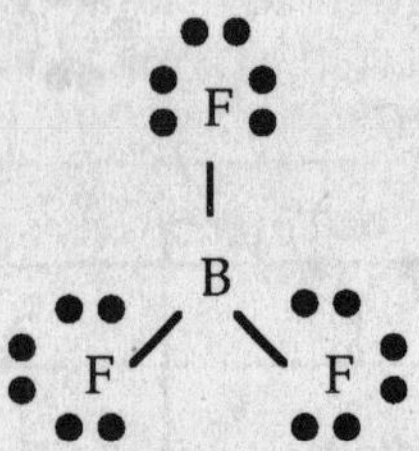

Central Atoms with More Than an Octet

Elements such as phosphorus (P) and sulfur (S) contain central atoms whose d-subshells are available at low enough energies to participate in bonding. In the case of PCl$_5$ the phosphorus has 10 electrons (5 bonds) surrounding it. The SF$_6$ has 12 electrons (6 bonds) around the sulfur.

PCl$_5$

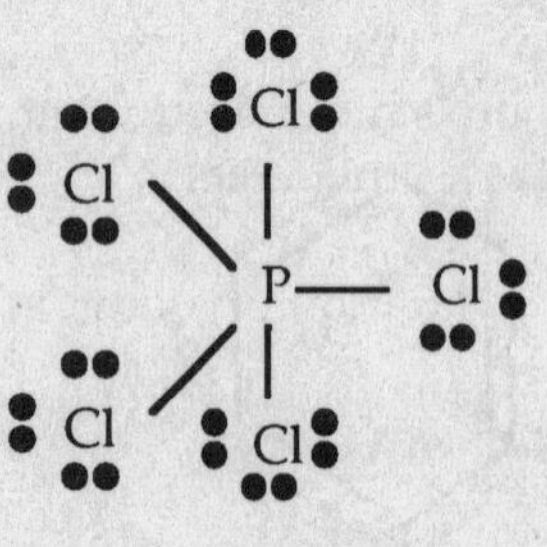

SF$_6$

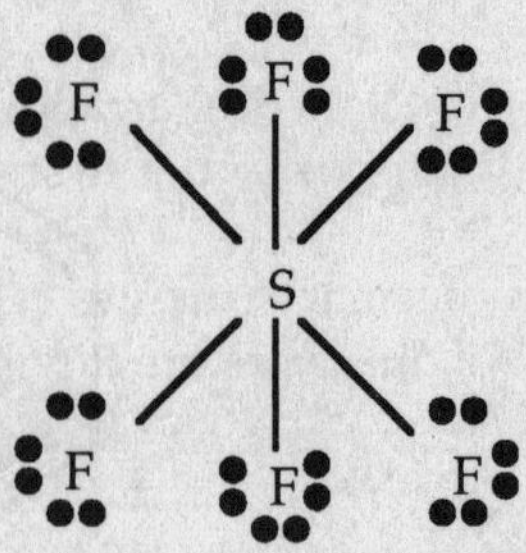

4. **Shapes of Molecules and Ions: Valence Shell Electron-Pair Repulsion Theory (VSEPR)**

The VSEPR model is a simple, effective way to predict the structure of molecules. These shapes play a very important role in predicting the chemical properties of nonmetallic compounds such as dipole moments. The main postulate of this model is that the bonding and nonbonding pairs around the central atom will be positioned as far from each other as is possible.

Use the following procedure to predict the structure of a molecule using the VSEPR theory. As you read the rules, confirm your understanding of each step by consulting the table of examples.

1. Write the Lewis structure of the molecule.

2. Use the Lewis structure to count the number of bonding pairs *and* lone pairs around the central atom. A lone pair is two electrons in the octet around the central atom which are not bonded to another atom.
An unpaired electron is counted the same as a lone pair. A single, double, or triple covalent bond is counted as *one* bonding electron region.

3. Identify the most stable arrangement of the electrons by counting the number of bonding regions and lone pairs around the central atom. Henceforth, bonding regions will be referred to simply as 'bonds'.

Total bonds *and* lone pairs	Shape
2	Linear
3	Trigonal planar
4	Tetrahedral
5	Trigonal bipyramidal
6	Octahedral

4. If more than one arrangement of lone pairs and chemical bonds is possible, select the one that will *minimize* lone pair repulsions.

    In trigonal bipyramidal arrangements, the repulsion is minimized when every lone pair is *in the plane of the triangle*.

    In octahedral arrangements the repulsion is minimized when the second lone pair is opposite the first *along an axis of the central atom*.

5. The molecular structure is determined by the location of atoms at the end of chemical bonds. Lone pairs are 'invisible'.

Example
The shape of molecules as a function of the number of bonding and nonbonding electron pairs

Number of electron pairs	Number of lone pairs	Shape of the bonds and lone pairs	Example	Geometry of the bonds	Basic bond angle
2	0	linear	$BeCl_2$	Cl—Be—Cl	180°
3	0	trigonal planar	BF_3	(trigonal planar with F, B, F, F)	120°
3	1	trigonal planar	SO_2	(S with two O, angular)	120°
4	0	tetrahedral	CCl_4	(tetrahedral C with 4 Cl)	109.5°
4	1	tetrahedral	NH_3	(trigonal pyramidal N with 3 H)	109.5° (See note.)
4	2	tetrahedral	H_2O	(bent or angular O with 2 H)	109.5° (See note.)
4	3	tetrahedral	HF	H—F	180°

Note: The actual bond angle of molecules such as NH_3 (107°) and H_2O (105°), which contain lone pair electrons, is less than the 'basic' bond angle of 109.5°. This is due to the repulsion of the bonding electrons by the lone pair electrons. One explanation attributes this effect to the larger geometrical area occupied by the lone pair electrons.

Number of electron pairs	Number of lone pairs	Shape of the bonds and lone pairs	Example	Geometry of the bonds	Basic Bond Angle
5	0	trigonal bipyramidal	PCl_5	trigonal bipyramidal	90° and 120°
5	1	trigonal bipyramidal	SF_4	irregular pyramidal	90° and 120° (See note on previous page.)
5	2	trigonal bipyramidal	ICl_3	T-shaped	90° and 120° (See note on previous page.)
5	3	trigonal bipyramidal	I_3^-	Linear	180°
6	0	octahedral	SF_6	octahedral	90°
6	1	octahedral	BrF_5	square pyramidal	90°
6	2	octahedral	XeF_4	square planar	90°

5. Attractive Forces

Molecules, atoms and ions are subject to electrostatic attractions of different strengths. These attractions can be caused by the electrical polarity between protons and electrons (metallic and covalent bonding), between positive and negative ions (ionic bonding), or between molecules (van der Waals forces). The emphasis in this chapter will be on the identification of the different attractions. In Chapter 4, States of Matter, the effect of the strength of attraction on physical properties will be reviewed.

Substances with the *strongest* attractive forces have a structure involving one of the three primary bonds between atoms or ions—metallic, ionic, and network covalent bonds.

- **Metallic solids:** The outer electrons of metal atoms are loosely held and move freely throughout these solids. The rest of the metal atom is a positive ion called the 'kernel', which occupies a fixed position in the crystal lattice. The strong attraction of the 'kernel' for the mobile electrons—the 'sea of moving electrons'—gives the metallic bond its strength. Metallic solids have high melting points, high boiling points, and high thermal and electrical conductivity in both the solid and liquid state.

- **Ionic solids:** When metals react with nonmetals, electrons are transferred from the metal to the nonmetal forming ions. The positive and negative ions that result have a strong attraction for each other and form an ionic solid. This attractive force is directly proportional to the charge on the ion and inversely proportional to the size of the ion—small ions such as Mg^{2+} and O^{2-} have more attraction for each other than the larger Na^+ and Cl^- ions. Ionic substances have high melting points, are hard and are brittle. They are excellent conductors of electricity when melted or in solution, but not as solids.

- **Network covalent solids:** Nonmetal atoms occupy the lattice positions in these solids and are bonded to each other by a three-dimensional *network* of covalent bonds. Network crystals have high melting points and are nonconductors of heat and electricity. Significant examples of these solids are carbon, C (graphite and diamond), silicon dioxide, SiO_2 (quartz and mica), silicon carbide, SiC, and boron nitride, BN.

Molecules are formed when nonmetal elements share electrons forming covalent bonds. The bond *between* molecules—the intermolecular attraction—is weak compared to the three primary bonds. This weak attraction between molecules is collectively referred to as a **van der Waals forces**. The force arises from the attraction of the nuclei on one molecule for the electrons of an adjacent molecule. Molecules have relatively *low* melting and boiling points. Molecules also exhibit little or no heat or electrical conductivity.

Intermolecular attractions are modeled on the concept of a dipole—a molecule with a positive and a negative end. There are three strengths for these dipoles, listed in descending order.

- **Hydrogen bonding:** Hydrogen bonding is the strongest intermolecular attractive force. Hydrogen bonding dipoles are formed when a molecule contains hydrogen bonded to a small, highly electronegative element—namely fluorine, F, oxygen, O, or nitrogen, N ("Happy summers are 'FON'"). Noteworthy examples are water, H_2O, ammonia, NH_3, and hydrogen fluoride, HF. A hydrogen bonding dipole is especially strong when compared to other intermolecular attractions. The result is a compound with an *unusually* high boiling and freezing temperature when compared to molecules with a similar molar mass or with a central atom from the same *group*.

Chemical Bonding

- **Dipoles**: Dipoles have an intermediate attraction. A molecular dipole—a polar covalent molecule—results when elements of different electronegativities bond covalently to form a *nonsymmetrical* molecule when the resulting molecule has both a positive (+) and a negative (–) end, it is a dipole.

  Dipoles are identified using the listed procedure.
  Ternary molecules are *always* dipoles. Examples are CH_3Cl, $HOCl$ and H_2SO_4
  Binary molecules *may* be dipoles.
  Binary molecules whose central atom is from group 5A, 6A, or 7A are dipoles. They have a nonsymmetrical shape. Examples are AsH_3, H_2S, and HCl.

- **Nonpolar molecules, London dispersion forces**: Nonpolar molecules have the least attraction for each other. A nonpolar covalent molecule results when elements of similar electronegativities bond covalently. The resulting molecule has neither a positive, +, nor a negative, – , end; it is a *not* a dipole. What little attraction the molecules have for each other is caused by a temporary, instantaneous dipole caused by the presence of more electrons on one side of the molecule than the other. The attraction between nonpolar molecules *increases* with an increase in the number of electrons in the molecule.

  Nonpolar molecules are identified using the listed procedure.
- **Mono-elemental** molecules are *always* nonpolar. Examples are H_2, O_2, N_2, F_2, Cl_2, Br_2, and I_2.
- **Binary** molecules *may* be nondipoles.
  Binary molecules whose central atom is from group 2A, 3A, or 4A are nonpolar molecules. They have a symmetrical shape. Examples are $BeCl_2$, BH_3, and CCl_4.

Examples

1. Which has the highest force of attraction between its molecules?
   - (A) hydrogen, H_2 (MM = 2.0)
   - (B) methane, CH_4 (MM = 16.0)
   - (C) nitrogen, N_2 (MM = 28.0)
   - (D) chlorine, Cl_2 (MM = 71.0)
   - (E) iodine, I_2 (MM = 254)

   Answer: **(E) iodine.**
   All of the choices are nonpolar molecules because they are mono-elemental molecules. Iodine has the highest number of electrons and therefore the greatest London dispersion force between its molecules.

2. Which has the strongest bonding in the solid state?
   - (A) chlorine, Cl_2
   - (B) hydrogen chloride, HCl
   - (C) oxygen, O_2
   - (D) sodium chloride, $NaCl$
   - (E) xenon, Xe

   Answer: **(D) sodium chloride.**
   Sodium chloride is an ionic compound, and is held together in the solid state by one of the three primary bonds. The other four are molecular, and are held together in the solid state by significantly weaker attractions—van der Waals forces.

The flow sheet shows the relationship between the six attractive forces.

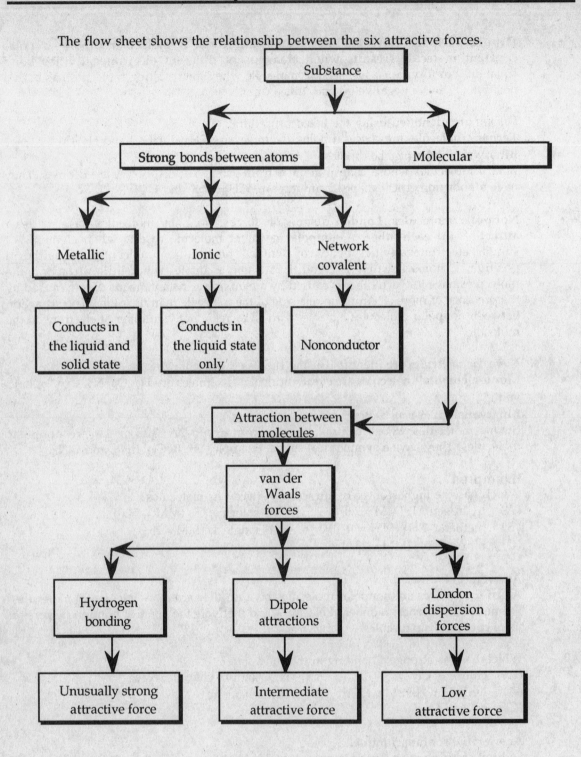

Chemical Bonding

6. Molecular Orbitals

The molecular orbital (MO) model was developed to predict electronic structure better than the resonance model. This theory handles molecules with unpaired electrons and with odd numbers of electrons better than the Lewis or VSEPR procedures. MO Theory allows correct estimates of the bond energy, the bond length and either paramagnetism (unpaired electrons) or diamagnetism (paired electrons) in molecules.

A set of molecular orbitals is formed from their corresponding atomic orbitals. These may be either bonding orbitals (between the atoms) or anti-bonding orbitals (either side of the atoms). A molecular orbital has energy, and the procedures used in determining the electron structure of atoms are followed here also. The 'aufbau' principle, the Pauli Exclusion Principle and Hund's Rule are used in describing molecular orbital structures.

Use the following procedure to predict the molecular orbital structure of a molecule. As you scan the examples, confirm your understanding of each step.

1. Count the electrons in the molecule or ion.

2. Electrons are assigned to the lowest-energy bonding (σ_{2s}, σ_{2p}, π_{2p}) or anti-bonding (σ_{2s}^*, σ_{2p}^*, π_{2p}^*) orbital available. Obey the Pauli Exclusion Principle (a maximum of 2 electrons per orbital) and Hund's Rule (electrons in equivalent energy orbitals occupy them singly first).

3. Bond 'order' = $\dfrac{\text{(no. of bonding electrons)} - \text{(no. of anti-bonding electrons)}}{2}$

4. Number of bonds = bond order

5. Bond order is directly proportional to bond energy, and inversely proportional to bond distance. A bond order of 3 (a triple covalent bond) represents a molecule with a large bond energy and a small bond distance.

6. **Diamagnetism** exists when all the electrons in the molecule are *paired*.
   Paramagnetism exists when there are one or more *unpaired* electrons in the molecule.

Examples: The molecular orbital structures for homonuclear species of the second period, namely beryllium, $_4Be_2$, boron, $_5B_2$, nitrogen, $_7N_2$, oxygen, $_8O_2$, and neon, $_{10}Ne_2$.

Note: There is a change in the order of filling that occurs when the structure of O_2 is written. Do not memorize the order of filling, but do be aware that it can change.

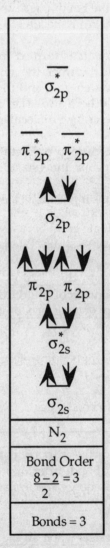
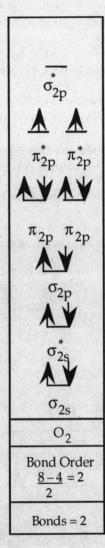

7. Hybrid Orbitals, Sigma Bonding, and Pi Bonding

A sigma (σ) bond occurs when two atomic orbitals combine to form a covalent bond that is along the axis connecting two atomic nuclei. Another covalent bond is the pi (π) bond. A pibond is a sausage-shaped region above and below the nuclei of bonded atoms. A double covalent bond consists of one σ and one π bond, and a triple covalent bond consists of one σ and two π bonds.

An important effect of a π bonding orbital is to fix the σ bond in a plane and to prevent rotation about a double-bond or a triple-bond.

A common occurrence during covalent bonding is the formation of sp, sp^2, or sp^3 'hybrid' atomic orbitals. To determine the kind of hybrid used in the bonding of the central atom in a molecule, count the number of bonds. A double or triple covalent bond counts as *one* bond. Two bonds result in an sp, three bonds gives an sp^2, and four bonds produces an sp^3 hybrid.

In this table, the Lewis structure is drawn for the molecule. The number of bonds is found by examining the bonding of each carbon atom, and the hybrid used by the carbon is determined by the number of bonds. The bond angle is set by the hybrid type. The number of sigma and pi bonds are then counted. The single bond between a hydrogen atom and a carbon atom counts as one sigma bond. The bonding between the carbon atoms accounts for the rest of the totals.

Structure	Number of bonds	Hybrid-ization	Bond angle	Sigma bonds	Pi bonds
H₃C—CH₃ (ethane)	4	sp^3	109.5°	7	0
H₂C=CH₂ (ethene)	3	sp^2	120°	5	1
H—C≡C—H	2	sp	180°	3	2

Example

Determine the electron hybridization of the central atom, the bond angles, and the number of sigma and pi bonds for CH_4, H_2CO, CO_2, and HCN.

Structure	Number of bonds	Hybrid- ization	Bond angle	Sigma bonds	Pi bonds
H–C(H)(H)–H (with H above and below)	4	sp^3	109.5°	4	0
H₂C=O	3	sp^2	120°	3	1
O=C=O	2	sp	180°	2	2
H–C≡N	2	sp	180°	2	2

8. Coordination Chemistry

Transition metal complexes are species in which ligands bond by contributing an electronpair to the central atom or ion. The naming of these complexes comes from an early model which generally assumed that coordinate covalent bonds are formed by involving the d-orbitals of the central metal ion. A complex ion is a distinct chemical, with properties different from its parts.

Many coordination compounds are chemically stable. The multiple oxidation states and electron structure of the transition metals often result in complexes which are vividly colored. The colors and stability make some complexes useful as confirmatory tests in schemes for determining the presence of ions in qualitative analysis.

Square planar and octahedral structures are the most common geometries for complexes. Familiarity with the systematic naming of these species permits insight into their use in analytical procedures and a practical use for chemicals with an octahedral structural geometry.

Procedure for naming coordination compounds

1. The cation (+ ion) is named before the anion.

2. The ligand is named before the metal.

3. End the name of a negative ligand with the letter 'o'.

Common Monodentate Ligands	
Ligand	Name
F^-, Cl^-, Br^-, I^-	Fluoro, chloro, bromo, iodo
:NO_2^-	Nitro
:CN^-	Cyano
:OH^-	Hydroxo
H_2O	Aquo
NH_3	Ammine

Common Polydentate Ligands			
Symbol	Ligand	Formula	Coordination Sites
en	Ethylenediamine	$NH_2-CH_2 = CH_2-NH_2$	2
EDTA	Ethylenediamine-tetraaceto		6

4. Count the number of ligands present by using Greek prefixes (mono, di, tri). If the ligand contains a Greek prefix (like ethylenediamine) use bis (for 2 of them) and tris (for 3 of them).

5. A Roman numeral shows the charge on the central metallic ion.

6. If the complex ion is negative, then the metal name ends with 'ate'.

7. The preferred order of naming ligands is alphabetical.

Examples

1.

	complex cation	anion
$[Cr(NH_3)_4Cl_2]Cl$	dichlorotetraamminechromium III	chloride
	$[2Cl^-]$ $[4\,NH_3]$ $[Cr^{3+}]$	$[Cl^-]$

Compounds with the same molecular formula but with different structures are called *isomers*. The diagram shows geometrical arrangements of ligands in two octahedral complexes with different physical and chemical properties.

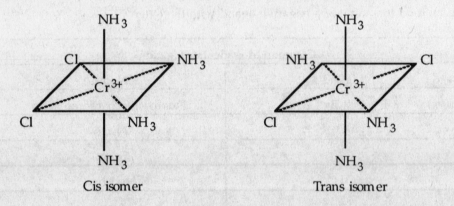

Cis isomer   Trans isomer

2.

	cation	complex anion
$K_2[PtCl_4]$	potassium	tetrachloroplatinate II
	$[K^+]$	$[4\,Cl^-]$ $[Pt^{2+}]$

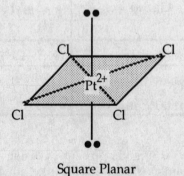

Square Planar

Chapter 3
Stoichiometry

1. **The Mole Concept and Molar Mass**

The mole is the amount of a substance that contains the same number of particles as are in exactly 12 grams of ^{12}C. This number is roughly 6 hecto (10^2) giga (10^9) tera (10^{12}) particles. The use of the mole concept in chemistry allows the grouping of large numbers of atoms, molecules, and ions instead of the impossible task of counting individual particles. A molar mass is the weight, in grams, of Avogadro's number of particles (6.022×10^{23}). Molar mass is computed by the use of atomic weights obtained from the periodic table of the elements.

Example: The mole concept.

1. A flask contains an unspecified number of krypton (Kr) atoms weighing 21.0 g. What weight of helium molecules contains an equal number of atoms?

   Moles of krypton molecules = $\dfrac{21.0 \text{ g}}{83.8 \text{ g mol}^{-1}}$ = 0.251 mol

   Moles of krypton molecules = moles of helium molecules
   Weight = 0.251 mol × 4.00 g mol^{-1} = 1.00 g

2. Which 1.0 g sample contains the greatest number of atoms?
   (A) Mercury, Hg

   moles of atoms = $\dfrac{1.0 \text{ g} \times 1 \text{ atom formula}^{-1}}{201 \text{ g mol}^{-1}}$ = 0.0050 mol

   (B) Carbon dioxide, CO_2

   moles of atoms = $\dfrac{1.0 \text{ g} \times 3 \text{ atoms molecule}^{-1}}{44.0 \text{ g mol}^{-1}}$ = 0.068 mol

   (C) Zinc sulfate, $ZnSO_4$

   moles of atoms = $\dfrac{1.0 \text{ g} \times 6 \text{ atoms formula}^{-1}}{162 \text{ g mol}^{-1}}$ = 0.037 mol

   (D) Butane, C_4H_{10}

   moles of atoms = $\dfrac{1.0 \text{ g} \times 14 \text{ atoms molecule}^{-1}}{58.0 \text{ g mol}^{-1}}$ = 0.24 mol

   (E) Hydrogen, H_2

   moles of atoms = $\dfrac{1.0 \text{ g} \times 2 \text{ atoms molecule}^{-1}}{2.0 \text{ g mol}^{-1}}$ = 1.0 mol

   Answer : (E)

The mole is a very convenient unit. The formula of a compound gives the relationship between the moles of each element of which it is comprised. Once the number of moles of a compound or the number of moles of one of the elements in a compound is known, the moles or grams of each of the parts of the compound can be computed. This is a 'mole link'.

The formula CO_2 can be read to mean that there are:

...twice as many moles of oxygen as moles of carbon. $n_{\text{oxygen atoms}} = \frac{2}{1} n_{\text{carbon atoms}}$

...half as many moles of carbon as moles of oxygen. $n_{\text{carbon atoms}} = \frac{1}{2} n_{\text{oxygen atoms}}$

...three times as many moles of atoms as moles of molecules. $n_{\text{atoms}} = \frac{3}{1} n_{\text{molecules}}$

Familiarity and comfort in the use of the 'mole link' will go a long way toward solving the many problems that use this concept. It is used to convert from moles of one element to moles of another in a compound, or the moles of the compound from the moles of one of its elements. Just remember to start the process by determining the moles of the substance whose identity and amount is given.

Example: 'Mole link' in formulas.

$$\text{Moles needed} = \frac{\text{needed}}{\text{known}} \times \text{Moles known}$$

1.

Formula	Needed	Known	Relationship
K_2SO_4	oxygen	potassium	$n_O = \frac{4}{2}n_K = 2n_K$
	sulfur	oxygen	$n_S = \frac{1}{4}n_O = \frac{1}{4}n_O$
	potassium	sulfur	$n_K = \frac{2}{1}n_S = 2n_S$
$Na_2C_8H_4O_4$	carbon	sodium	$n_C = \frac{8}{2}n_{Na} = 4n_{Na}$
	sodium	hydrogen	$n_{Na} = \frac{2}{4}n_H = \frac{1}{2}n_H$
	oxygen	carbon	$n_O = \frac{4}{8}n_C = \frac{1}{2}n_C$

2. Element X reacts with oxygen to produce a pure sample of X_4O_{10}. Find the atomic weight and the identity of X if an experiment reacting oxygen with 1.00 g of X resulted in 2.29 of X_4O_{10}.

$$\text{Moles of oxygen} = \frac{(2.29 - 1.00)\,g}{16.0\,g\,mole^{-1}} = \frac{1.29\,g}{16.0\,g\,mole^{-1}} = 0.0806\,\text{moles}$$

$n_X = \frac{4}{10}n_O = \frac{2}{5} \times 0.0806 = 0.0323$ mole

Molar mass $\quad = \quad \dfrac{1.000\,g}{0.0323\,mol} \quad = \quad 31.0\,g\,mol^{-1}$

The element could be phosphorus, P.

2. **Empirical Formulas and Molecular Formulas**

The ratio of combined atoms in a given compound cannot be changed. To do so would be to describe a different substance. The simplest whole-number ratio of atoms is the 'empirical formula'.

The procedure to compute the empirical formula is:
(a) If the composition (either weight or weight percent) of a compound is known, the moles of each element is calculated.
(b) The mole ratio is determined by dividing the moles of each element of the compound by the smallest number of moles.
(c) An adjustment may be required to adjust the ratio to a whole-number ratio (empirical formula).

A 'molecular formula' is a simple integer number, the multiple (including one) of an empirical formula. The molar mass is required in the computation, and the multiple is the result of dividing the molar mass by the weight of the empirical formula.

Example

1. Analysis of an organic compound reveals the composition by mass: 22.79% carbon, 1.42% hydrogen, and 75.79% bromine. The molar mass is determined to be 421.7 g mol^{-1}.

   What are the empirical and molecular formula of the compound?

   (a) Calculate the empirical formula.

Element	Weight	Moles	Ratio	Whole #
carbon	22.79 g	1.90	2.00	4
hydrogen	1.42 g	1.42	1.50	3
bromine	75.79 g	0.948	1.00	2

   (b) The empirical formula is $C_4H_3Br_2$.

   (c) The empirical weight is 210.8 g mol^{-1}.

   (d) The molar mass, 421.7 g mol^{-1}, is twice the empirical weight: (421.7 ÷ 210.8 = 2.000).

   (e) The molecular formula is $(C_4H_3Br_2)_2$ or $C_8H_6Br_4$.

2. When 1.000 g of an organic compound with the formula $C_xH_yO_z$ is burned in excess oxygen, 1.667 g of carbon dioxide (CO_2), and 0.545 g of water vapor (H_2O) are obtained.

   What is the empirical formula of the compound?

   (a) $n_{carbon\ dioxide} = \dfrac{1.667\ g}{44.0\ g\ mol^{-1}} = 0.0379\ mol$

   (b) $n_{water\ vapor} = \dfrac{0.545\ g}{18.0\ g\ mol^{-1}} = 0.0303\ mol$

   (c) The moles of carbon are equal to the moles of CO_2.
   $n_C = \frac{1}{1} n_{CO_2} = 1 \times 0.0379 = 0.0379\ mol$

   (d) The moles of hydrogen are twice the moles of H_2O.
   $n_H = \frac{2}{1} n_{H_2O} = 2 \times 0.0303 = 0.0606\ mol$

   (e) The weights of carbon and hydrogen in the compound are computed by multiplying the moles by the molar mass.

   $wt_C = 0.0379\ mol \times 12.0\ g\ mol^{-1} = 0.455\ g$

   $wt_H = 0.0606\ mol \times 1.0\ g\ mol^{-1} = 0.061\ g$

(f) The weight of oxygen is found by adding the weight of carbon and hydrogen, and then subtracting this from the total weight of starting compound, 1.000 g
$wt_{oxygen} = 1.000 - (0.455 + 0.061) = 0.484$ g
The moles of oxygen are then computed.

Element	Moles$_1$	Weight	Moles$_2$	Ratio	Whole #
carbon	0.0379	0.455 g	0.0379	1.25	5
hydrogen	0.0606	0.061 g	0.0606	2.00	8
oxygen		0.484 g	0.0303	1.00	4
Total		1.000 g			

Answer: The empirical formula would be $C_5H_8O_4$.

3. **Reaction Stoichiometry**

The coefficients in a balanced chemical reaction equation are directly proportional to the number of moles. The equation

$$H_2(g) + Cl_2(g) \Rightarrow 2HCl(g)$$

can be read to say: "One mole of hydrogen, H_2, reacts with one mole of chlorine, Cl_2, to give two moles of hydrogen chloride, HCl."

There is a 'mole link' between the substances in a reaction and this permits the use of a procedure similar to that used in the last section to determine the relationship between the elements in a compound. The 'mole link' allows the weight or the moles of any substance in a chemical equation to be computed when either the weight or the moles of any other chemical in the reaction is given.

This type of problem assumes that:
(a) The reaction proceeds 100% to completion.
(b) There is an 'excess' of the other reactant to produce an amount predicted by consuming the weight of the known substance.

Example

When 9.53 g of carbon disulfide is reacted with an excess of oxygen, carbon dioxide, and sulfur dioxide are formed. Find the mass of sulfur dioxide formed in this reaction.

(a) Write and balance the reaction equation.
$CS_2(g) + 3O_2(g) \Rightarrow CO_2(g) + 2SO_2(g)$
(b) Compute the moles of the given substance.
$n_{carbon\ disulfide} = \dfrac{9.53\ g}{76.2\ g\ mol^{-1}} = 0.125$ mol
(c) Use the 'mole link' to determine the moles of the product of the reaction.
$n_{sulfur\ dioxide} = \frac{2}{1} n_{carbon\ disulfide} = 2 \times 0.125$
$n_{sulfur\ dioxide} = 0.250$ mol
(d) Determine the weight of a substance.
$wt_{sulfur\ dioxide} = 0.250$ mol $\times 64.0$ g mol^{-1} = **16.0 g**

4. Limiting Reactants

Limiting reaction problems are where the quantities of *two* (rarely more than two) reactants are stated. The problem allows the assumption of 100% reaction for one of the substances. It is then necessary to determine which of the two would be used up...the 'limiting' reactant. The 'limiting' reactant is then used with the 'mole link' to compute the amount of product produced.

Example

The reaction of chromium with sulfur produces chromium(III) sulfide. How many grams of chromium(III) sulfide can be isolated when 26.0 g of chromium is reacted with 77.0 g of sulfur?

(a) Write and balance the reaction equation.
$16Cr(s) + 3S_8(s) \Rightarrow 8Cr_2S_3(s)$

(b) Compute the moles of the given substances. In any problem, this leads to a rule.
When in doubt about a problem type, start the solution of the problem by computing moles using the amounts and molar masses available.

$n_{chromium} = \dfrac{26.0 \text{ g}}{52.0 \text{ g mol}^{-1}} = 0.500 \text{ mol}$

$n_{sulfur} = \dfrac{77.0 \text{ g}}{257 \text{ g mol}^{-1}} = 0.300 \text{ mol}$

(c) Use the 'mole link' to determine the 'limiting reactant'. At least one of the reactants will be completely consumed in this type of problem.

$n_{chromium} = \tfrac{16}{3} n_{sulfur} = \tfrac{16}{3} \times 0.300 = 1.60 \text{ mol}$
This exceeds the amount of chromium.
Chromium is the 'limiting reactant'.

$n_{sulfur} = \tfrac{3}{16} n_{chromium} = \tfrac{3}{16} \times 0.500 = 0.0938 \text{ mol}$
This does not exceed or equal the amount of sulfur given.
Sulfur is *not* the 'limiting reactant'.

(d) Determine the weight of the product.

$n_{chromium\ sulfide} = \tfrac{8}{16} n_{chromium} = \tfrac{1}{2} \times 0.500$
$n_{chromium\ sulfide} = 0.250 \text{ mol}$

$wt_{chromium\ sulfide} = 0.250 \text{ mol} \times 200. \text{ g mol}^{-1}$
$wt_{chromium\ sulfide} = 50.0 \text{ g}$

5. Molarity, Molality, and Mole Fraction

A solution is a homogeneous mixture of one or more solutes and a solvent.

(Solute + Solvent = Solution)

Many reactions occur in solution, particularly in aqueous (water) solution. The most common concentration used in chemistry to indicate the quantity of solute dissolved in a given amount of solvent is molarity.

$$\text{Molarity (M)} = \frac{\text{moles solute (n)}}{\text{Liter solution (V)}}$$

The freezing point depression and the boiling point elevation measurements of a solution is frequently used to determine the molar mass or the percent ionization of a solute. The temperature change depends upon the number of molecules or ions dissolved in a specified amount of solvent, and the preferred concentration unit is molality.

$$\text{Molality (m)} = \frac{\text{moles solute (n)}}{\text{kilogram solvent (kg)}}$$

Solutions of a nonvolatile solute in a liquid solvent will have a lower vapor pressure than the pure solvent. Raoult's Law allows the calculation of the vapor pressure of a solution by multiplying concentration by the vapor pressure of the solvent. The concentration unit used is mole fraction.

$$\text{Mole fraction (x)} = \frac{\text{moles solute}}{\text{mole solution}}$$

Note that:

(a) The numerator in the three concentration expressions is the same. Converting from one to another requires adjusting the denominator.

(b) The *density* of the solution is used to convert between molarity and molality. The *molar mass* of the solvent is used to convert between mole fraction and molality.

(c) It is inconvenient to change between molarity and mole fraction directly, so molality should be calculated as an intermediate step.

$$\text{Molarity (M)} \quad \xleftrightarrow{\text{density}} \quad \text{Molality (m)} \quad \xleftrightarrow{\text{MM}_{\text{solvent}}} \quad \text{Mole fraction (x)}$$

(d) For the same solution, molality always has the greatest numerical value, and mole fraction the lowest.

$$\text{Mole fraction (x)} \quad << \quad \text{Molarity (M)} \quad < \quad \text{Molality (m)}$$

Examples

1. The density of a solution of 20.0% (by weight) KI (MM = 166.03) solution in water is 1.166 g mL^{-1} (1.166 kg L^{-1}).

   $$1000 \text{ g}_{solution} = 200 \text{ g}_{solute} + 800 \text{ g}_{solvent}$$
   $$1.000 \text{ kg}_{solution} = 0.200 \text{ kg}_{solute} + 0.800 \text{ kg}_{solvent}$$

   (a) Calculate the molarity.

   $$n_{solute} = \frac{200 \text{ g}}{166 \text{ g mol}^{-1}} = 1.20 \text{ mol}$$

   $$L_{solution} = \frac{1.000 \text{ kg}_{solution}}{1.166 \text{ kg L}^{-1}} = 0.858 \text{ L}$$

   $$\text{Molarity} = \frac{1.20 \text{ mol}}{0.858 \text{ L}} = \mathbf{1.40 \text{ M}}$$

   (b) Calculate the molality.

   $$n_{solute} = 1.20 \text{ mol} \quad \text{(from part (a))}$$
   $$kg_{solvent} = 0.800 \text{ kg}$$
   $$\text{Molality} = \frac{1.20 \text{ mol}}{0.800 \text{ kg}} = \mathbf{1.50 \text{ m}}$$

   (c) Calculate the mole fraction.

   $$n_{solute} = 1.20 \text{ mol} \quad \text{(from part (a))}$$
   $$n_{solvent} = \frac{800. \text{ g}_{solvent}}{18.0 \text{ g mol}^{-1}} = 44.4 \text{ mol}$$

   $$n_{solution} = 44.4 \text{ mol}_{solvent} + 1.20 \text{ mol}_{solute} = 45.6 \text{ mol}$$

   $$\text{mole frac.} = \frac{1.20 \text{ mol}}{45.6 \text{ mol}} = \mathbf{0.0263}$$

2. A 1.00 molal solution of NaOH(aq) has a density of 1.032 g mL^{-1}.

   (a) Calculate the molarity.

   $$n_{solute} = 1.00 \text{ mol} \quad \text{(from 1.00 molal)}$$

   $$L_{solution} = \frac{\left(1.000 \text{ kg}_{solvent} + 0.040 \text{ kg}_{solute}\right)}{1.032 \text{ kg L}^{-1}} = 1.01 \text{ L}$$

   $$\text{Molarity} = \frac{1.00 \text{ mol}}{1.01 \text{ L}} = \mathbf{0.992 \text{ M}}$$

   (b) Calculate the mole fraction.

   $$n_{solute} = 1.00 \text{ mol} \quad \text{(from 1.00 molal)}$$
   $$n_{solvent} = \frac{1000 \text{ g}_{solvent}}{18.0 \text{ g mol}^{-1}} = 55.6 \text{ mol}$$

   $$n_{solution} = 55.6 \text{ mol}_{solvent} + 1.00 \text{ mol}_{solute} = 56.6 \text{ mol}$$

   $$\text{mole frac.} = \frac{1.00 \text{ mol}_{solute}}{55.6 \text{ mol}_{solution}} = \mathbf{0.0180}$$

6. Solution Stoichiometry

The labels on reagent bottles gives the molarity (M) of each solution used in a reaction. The moles of solute are calculated using the molarity and volume of each solution in the reaction (moles = Molarity x Volume). The coefficients of the balanced reaction equation provide a 'mole link' which can be used to compute the moles of a substance that react with are produced from a known amount of a reactant.

Example

When 20.0 mL of 0.15 M magnesium sulfate is reacted with a 0.10 M sodium hydroxide solution, magnesium hydroxide is precipitated. Find the minimum amount of NaOH(aq) required to completely react with the $MgSO_4$(aq); and the maximum yield of dried precipitate that would be expected.

(a) Write and balance the reaction equation.
$$Mg^{2+}(aq) + 2OH^-(aq) \Rightarrow Mg(OH)_2(s)$$

(b) Compute the moles of the given substance.
$$n_{Mg^{2+}} = 0.15 \text{ mol L}^{-1} \times 0.020 \text{ L} = 0.0030 \text{ mol}$$

(c) Use the 'mole link' to determine the moles of the reactant required.
$$n_{OH^-} = {}^2/_1 \, n_{Mg^{2+}} = 2 \times 0.0030$$
$$n_{OH^-} = 0.0060 \text{ mol}$$
$$n_{Mg(OH)_2} = n_{Mg^{2+}} = 0.0030 \text{ mol}$$

(d) Determine the volume of reactant needed and the weight of product that will be produced.
$$mL_{NaOH} = \frac{0.0060 \text{ mol}}{0.10 \text{ mol L}^{-1}} = 0.060 \text{ L} = 60. \text{ mL}$$
$$wt_{Mg(OH)_2} = 0.0030 \text{ mol} \times 58.3 \text{ g mol}^{-1} = \mathbf{1.75 \text{ g}}$$

7. Colligative Properties

Freezing point depression (ΔT_f), boiling point elevation (ΔT_b), and vapor pressure lowering ($\Delta P°$) depend on the number of solute particles, ions or molecules, present in a solution. A lower vapor pressure is caused by the attraction between solute and solvent molecules. The freezing point and boiling point of the solution are dependent upon vapor pressure. The result is a *depression* of the freezing point and an *elevation* of the boiling point of the solution when compared to those of the solvent. The concentration unit of choice used in colligative property equations is either molality (m) or mole fraction (x).

The symbols for the molal freezing and boiling point constants are K_f and K_b, and these values are *always* supplied. The symbol $P°$ is used for the vapor pressure of the pure solvent, and it too is always supplied.

The key formulas are:

fp and bp change	$\Delta T_f = K_f m$	$\Delta T_b = K_b m$
Raoult's Law	$P_{solvent} = x_{solvent} P°$	$\Delta P_{solvent} = x_{solute} P°$

Example

1. In a determination of a melting point to approximate the molar mass of a compound, 1.005 g of the compound is dissolved in 12.07 g of naphthalene. The freezing point was lowered by 4.4°C. The freezing point constant (K_f) for naphthalene is 6.6° molal^{-1}. What is the approximate molar mass of the compound?

   (a) Determine the weight of solute per kg solvent.

   $$g_{solute} \, kg^{-1} = \frac{1.005 \, g_{solute}}{12.07 \, g_{solvent}} \times 1000 \, g \, kg^{-1}$$

   $$g_{solute} \, kg^{-1} = 83.3 \, g \, kg^{-1}$$

   (b) Determine the molality of the solution.

   $$\text{molality} = \frac{\Delta T_f}{K_f} = \frac{4.4 \, °C}{6.6 \, °C \, m^{-1}} = 0.67 \, mol \, kg^{-1}$$

   (c) Determine the molar mass of the solute.

   $$MM = \frac{83.3 \, g \, kg^{-1}}{0.67 \, mol \, kg^{-1}} = 120 \, g \, mol^{-1}$$

2. The vapor pressure of toluene is 28.829 mmHg. A solution of a nonvolatile, nonionizing solute dissolved in toluene is made. When 4.256 g compound is dissolved in 100.0 g of toluene (C_7H_8, MM = 92.07) the vapor pressure becomes 28.209 mmHg. What is the approximate molar mass, MM, of the solute?

   (a) Find the mole fraction of solvent using Raoult's Law.

   $$x_{solute} = \frac{\Delta P_{solvent}}{P°_{solvent}} = \frac{(28.829 - 28.209)}{28.829} = 0.0215$$

   $$x_{solvent} = 1.0000 - 0.0215 = 0.9785$$

   (b) Determine the moles of solvent, solute and solution.

   $$n_{solvent} = \frac{100.0 \, g}{92.07 \, g \, mol^{-1}} = 1.086 \, mol$$

   $$n_{solute} = \frac{4.256 \, g}{MM \, g \, mol^{-1}} = 4.256 \div MM$$

   $$n_{solution} = 1.086 + (4.256 \div MM)$$

   (c) Calculate molar mass using the mole fraction definition.

   $$x_{solvent} = 0.9785 = \frac{1.086}{1.086 + (4.256 \div MM)}$$

   $$MM = 178.0 \, g \, mol^{-1}$$

Chapter 4
States of Matter

1. The Solid, Liquid, and Gas States

In the process of melting or sublimation, the atoms, ions, or molecules which make up a solid become separated. The more strongly the units of a lattice attract each other, the greater is the temperature required to separate them.

High melting crystalline solids are held together by *bonds*-either metallic, ionic, or network covalent. The few substances that form network covalent bonds include carbon in the form of diamond or graphite and silicon dioxide (SiO_2) in the from of quartz, mica, or asbestos.

Molecules are held in crystal lattices by *intermolecular attractive forces*. The melting temperatures of these substances are significantly lower than those of metals and salts. The boiling temperature of molecules are generally lower than those of metallic, ionic, and network substances since molecular substances have significantly higher vapor pressures at similar temperatures.

Hydrogen bonding is the strongest intermolecular force of attraction. This type of attraction always involves molecules containing hydrogen atoms, H sharing an electron-pair with a small, highly electronegative atom—either nitrogen, N, oxygen. O, or fluorine, F. The hydrogen nucleus is both electron-deficient and unshielded by inner-shell electrons. It makes up for this electron deficiency by sharing the lone-pair electrons on a nearby molecule containing nitrogen, oxygen, or fluorine.

Example: Hydrogen bonding in hydrogen fluoride, HF.

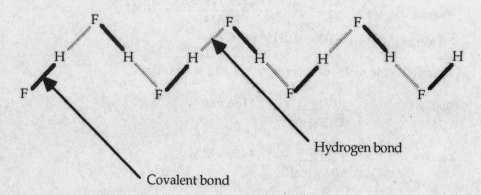

Van der Waals forces are the weakest attractive forces between molecules. There are two major van der Waals' forces.

1. **Dipole attractions** occur when polar molecules are attracted to one another.

2. **Dispersion forces** are the weakest of all molecular attractions, and are caused by instantaneous, temporary dipoles caused by the motion of electrons. The strength of dispersion forces increases as the number of electrons in a molecule increases. This is now called a 'London dispersion force'.

Example

State which has an ionic, a molecular covalent, or a network covalent structure.

	Formula	Data and properties		Structure	Explanation
1.	HBr	mp $-88.5°$C	bp $-67°$C	**Molecular Dipole**	Nonsymmetrical binary compound (central atom from Group 7A)
2.	BN	Sublimes at $3000°$C All single bonds		**Network Covalent Molecule**	Very high sublimation temperature
3.	CS_2	mp $-110.8°$C	bp $46.3°$C	**Molecular Nondipole**	Symmetrical binary compound (central atom from Group 4A)
		Double covalent bonds between C=S			
4.	$MgCl_2$	mp $708°$C	bp $1412°$ C	**Ionic Salt**	Very high melting temperature
		Conducts electricity when molten			
5.	$AsCl_3$	mp $-8.5°$C	bp $130.2°$C	**Molecular Dipole**	Nonsymmetrical binary compound (central atom from Group 5A)

2. Changes of State

A graph of pressure plotted against temperature displays the regions of all the phases of a pure substance at the same time. The phase change diagram for benzene, C_6H_6, shows several characteristics which can be illustrated by this type of presentation.

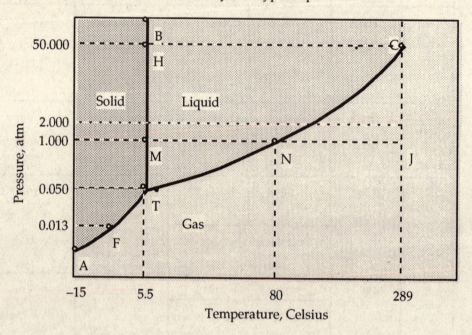

The equilibrium **vapor pressure** is the pressure of the vapor in equilibrium with a liquid or solid at a given temperature. It is dependent upon temperature only, and is represented in the diagram by segments 'TC' and 'AT'.

Examples: Properties illustrated in a phase change diagram.

1. The **sublimation vapor pressure** curve is segment AT.

    Sublimation is the vaporization of a solid. Solids have an equilibrium vapor pressure at any given temperature. Molecular solids are fairly volatile, which is why a paraffin candle has an odor.
    Benzene at a pressure of 10 mmHg (0.013 atm) will sublime (change from the solid state directly into the gaseous state) at a temperature corresponding to point 'F'—about 11°C.

2. The **vapor pressure** curve is TC.

    Liquids have an equilibrium vapor pressure at any given temperature. The vapor pressure of a selected substance is dependent *on temperature alone.* Molecular liquids are fairly volatile, which is why water, alcohol, C_2H_5OH, and acetone, CH_3COCH_3, vaporize or evaporate easily.
    The *normal* boiling temperature is the temperature at which the vapor pressure of a substance is equal to 1.00 atmosphere (760 mmHg).
    At a pressure of 1 atmosphere, benzene will boil at the temperature corresponding to point 'N'(80°C).

3. The **melting temperature** curve of benzene is TB.

The melting temperature curve for benzene is nearly vertical, and the melting point of this substance shows little change with pressure. The melting temperature curve for water, H_2O, has a negative slope (\), and the melting point of H_2O *decreases* with an increase in pressure. Carbon dioxide, CO_2, has a melting temperature curve with a positive slope (/), which is most common. The melting point of CO_2 *increases* with an increase in pressure.

The normal melting temperature of benzene is 5.5°C. At a pressure of 1 atmosphere, benzene will melt at the temperature corresponding to point 'M' (5.5°C).

4. The **triple point** of benzene is 'T'.

At the triple point all three phases—solid, liquid, and gas—are in equilibrium. At a pressure of 36 mmHg (0.05 atm) and a temperature of 5.5°C all three states for benzene—solid, liquid, and gas—exist simultaneously.

5. The **critical point** of benzene is 'C'.

The *critical temperature*, is the temperature above which it is impossible to liquefy a gas no matter how much the pressure is increased. The *critical temperature* of benzene is 289°C.

The critical point also marks the *critical pressure*, which is the minimum amount of pressure required to liquefy a gas at the critical temperature. The graph shows that the critical pressure of benzene is 50 atm.

3. Kinetic Molecular Theory of Gases; Graham's Law

The kinetic theory of gases postulates that gaseous molecules move in a straight-line at high velocities.

(1) The collision of the molecules with the walls of a container causes gaseous pressure. Decreasing the volume of the vessel will cause the pressure to increase, since there are more collisions per square unit of wall area (Boyle's Law).

(2) The average kinetic energy (mass × velocity squared, mv^2) is proportional to the temperature. Increasing the temperature of the gas will cause the pressure to increase, since the molecules are moving faster and hitting the wall at an increased number of collisions per unit time (Charles' Law).

Gases diffuse or mix, and the rate at which this occurs is dependent on the speed of the gas molecules, which in turn is dependent upon the temperature. The escape of a gas through a pin hole is called **effusion**.

Graham's Law relates the rate of effusion of two different gases at the same temperature (the same average kinetic energy). The rate of effusion and the rate of diffusion of two gases is close enough so that the terms are used interchangeably.

$$\frac{\text{Rate of effusion of Gas A}}{\text{Rate of effusion of Gas B}} = \sqrt{\frac{\text{Molar Mass of Gas B}}{\text{Molar Mass of Gas A}}}$$

Example

1. What is the relative rate of effusion of CO (MM = 28.0) and CO_2 (MM = 44.0)?
   - (A) CO is 1.25 times faster than CO_2
   - (B) CO_2 is 1.25 times faster than CO
   - (C) CO is 1.57 times faster than CO_2
   - (D) CO_2 is 1.57 times faster than CO
   - (E) They diffuse at the same rate

   The answer is (**A**).

$$\frac{\text{Rate of effusion of CO}}{\text{Rate of effusion of CO}_2} = \sqrt{\frac{\text{Molar Mass of CO}_2}{\text{Molar Mass of CO}}} = \sqrt{\frac{44.0}{28.0}}$$

$$\frac{\text{Rate of effusion of CO}}{\text{Rate of effusion of CO}_2} = 1.25$$

2. Which of these gases effuses more quickly than oxygen, O_2?
   - (A) F_2
   - (B) N_2O
   - (C) H_2S
   - (D) NO
   - (E) Cl_2

   The answer is (**D**).
   NO (MM = 30) is the only choice which has a lower molar mass than oxygen (MM = 32).

4. The Ideal Gas Law

Any amount of a gas will occupy the volume of the rigid vessel in which it is contained no matter what the pressure or temperature might be. The *amount* (n, in moles), the *volume* (V, in liters), the *pressure* (P, in atmospheres), and the *temperature* (T, in kelvins) are related by the Ideal Gas Law Equation. R is the universal gas constant. When any three of the variables in the equation are known, the fourth can be computed.

$$PV = nRT$$

Real gases differ from ideal gases in two ways.

(1) Real gases are subject to van der Waals forces of attraction. In calculations requiring high accuracy, a term $\frac{n^2a}{V^2}$ is added to the pressure, $\left(P + \frac{n^2a}{V^2}\right)$, to correct for the attractions between molecules.

(2) Real gas molecules take up space. The term 'nb' is subtracted from the volume, (V − nb), to correct for that portion of the volume that is not compressible because real gases have an intrinsic volume.

Examples

Rearrangements of the ideal gas law equation allow its use to calculate gaseous density or molar mass from experimental data.

(1) To calculate **gas density**, D:

$$PV = [n]RT$$

$$n = \frac{g}{MM}$$

$$PV = \left[\frac{g}{MM}\right]RT$$

$$D = \frac{g}{V} = \frac{P \times MM}{RT}$$

(2) To calculate **molar mass**, MM:

$$PV = [n]RT$$

$$n = \frac{g}{MM}$$

$$PV = \left[\frac{g}{MM}\right]RT$$

$$MM = \frac{g \times R \times T}{P \times V}$$

1. What is the *density* of nitrogen gas (g L^{-1}) in a sample that exerts a pressure of 150. mmHg at 0°C?

$$D = \frac{g}{V} = \frac{P \times MM}{RT} = \frac{\left(\frac{150}{760}\right) \times 28.0 \text{ g mol}^{-1}}{\left(0.0821 \frac{\text{L atm}}{\text{mol K}}\right) \times 273 \text{ K}} = 0.247 \text{ g L}^{-1}$$

2. 1.135 g of a gaseous compound containing only nitrogen and oxygen occupies 642 mL at a pressure of 720 mmHg and a temperature of 27.0°C. What is the *molar mass* of this gas?

$$MM = \frac{g \times R \times T}{P \times V} = \frac{1.135 \text{ g} \times 0.0821 \frac{\text{L atm}}{\text{mol K}} \times 300 \text{ K}}{\left(\frac{720}{760}\right) \text{atm} \times \left(\frac{642}{1000}\right) \text{L}} = 46.0 \text{ g mol}^{-1}$$

5. Dalton's Law of Partial Pressures

Each gas in a mixture exerts a partial pressure as if it were present alone. This pressure is proportional to the number of moles of the gas present.

Example

An equimolar mixture of methane, CH_4, and ethyne, C_2H_2, occupies a certain volume at a total pressure of 100. mmHg. The sample is burned, and the CO_2 alone is collected. The pressure of the CO_2 is found to be 150. mmHg at the same temperature as the original mixture. What fraction of the gas mixture is CH_4?

$$CH_4(g) + 2O_2(g) \Rightarrow CO_2(g) + H_2O(l)$$
$$C_2H_2(g) + \tfrac{5}{2} O_2(g) \Rightarrow 2CO_2(g) + H_2O(l)$$

Of the CO_2 that results from the combustion, ⅓ comes from the methane, CH_4, and ⅔ comes from the ethene, C_2H_2.

Final mixture:
⅓ × 150. = 50.0 mmHg is the partial pressure of the CO_2 resulting from the CH_4
⅔ × 150. = 100. mmHg is the partial pressure of the CO_2 resulting from the C_2H_2

Initial mixture:

The partial pressure of the CH_4 = $\dfrac{1 \text{ mol } CH_4}{1 \text{ mol } CO_2} \times 50.0$ = 50.0 mmHg

The partial pressure of the C_2H_2 = $\dfrac{1 \text{ mol } C_2H_2}{2 \text{ mol } CO_2} \times 100.$ = 50.0 mmHg

The fraction of the gas that is CH_4 is ½ or 0.50.

6. Properties of Solutions; Raoult's Law

The decrease in the vapor pressure of a solution—the vapor pressure of the solution is *lower* than that of the pure solvent—is proportional to the mole fraction of the nonvolatile solute dissolved in the solvent.

$$\Delta P = x_{solute} P°_{solvent}$$

where x_{solute} is the mole fraction of solute,
and $P°_{solvent}$ is the vapor pressure of the pure *solvent*

When the solution is made from *two* volatile liquids, the partial pressure of each of the liquids in the vapor above the solution is calculated from Raoult's Law.

$$P_{component} = x_{component} P°_{component}$$

where $P_{component}$ is the vapor pressure of the liquid in the solution mixture,
and $x_{component}$ is the mole fraction of the liquid,
and $P°_{component}$ is the vapor pressure of the pure *liquid*

$$P_{component_1} = x_{component_1} P°_{component_1}$$
$$P_{component_2} = x_{component_2} P°_{component_2}$$

and the total pressure of the vapor is derived from Dalton's Law of Partial Pressures.

$$P_{Total} = P_{component_1} + P_{component_2}$$
$$P_{Total} = x_{component_1} P°_{component_1} + x_{component_2} P°_{component_2}$$

States of Matter

Examples

1. What is the vapor pressure at 34.9°C of ethanol in a solution prepared by mixing 184 g of ethanol, C_2H_5OH (MM = 46.0) is dissolved in 92.1 g of glycerol, $C_3H_5(OH)_3$ (MM = 92.1). The vapor pressure of pure ethanol at 34.9°C is 100.0 mmHg and glycerol is nonvolatile at 34.9°C.

$$x_{ethanol} = \frac{\left(\dfrac{184.0\ g}{46.0\ g\ mol^{-1}}\right)}{\left(\dfrac{92.1\ g}{92.1\ g\ mol^{-1}}\right) + \left(\dfrac{184.0\ g}{46.0\ g\ mol^{-1}}\right)} = 0.800$$

$P_{ethanol} = x_{ethanol}\ P°_{ethanol} = 0.800 \times 100.0\ mmHg =$ **80.0 mmHg**

2. The vapor pressure of benzene, C_6H_6, and toluene, C_7H_8, are 95.1 mmHg and 28.4 mmHg, respectively. What is the composition of the vapor *above* an equimolar ($x_{benzene} = x_{toluene} = 0.500$) solution of the two volatile components?

    (a) Calculate the total pressure exerted by the solution.

$P_{benzene}$	$P_{toluene}$
$= x_{benzene}\ P°_{benzene}$	$= x_{toluene}\ P°_{toluene}$
$= 0.500 \times 95.1$	$= 0.500 \times 28.4$
$= 47.6$ mmHg	$= 14.2$ mmHg
$P_{Total} = 47.6 + 14.2 =$ **61.8 mmHg**	

    (b) Dalton's law of partial pressures predicts that the ratio of the partial pressure of each component to the total pressure is equal to the mole fraction of that component.

$x_{benzene}$	$x_{toluene}$
$= P_{benzene} \div P_{Total}$	$= P_{toluene} \div P_{Total}$
$= 47.6 \div 61.8 = 0.770$	$= 14.2 \div 61.8 = 0.230$
Answer: The vapor is **77.0% benzene** and **23.0% toluene** by pressure and by moles.	

Chapter 5
Reaction Kinetics

1. **Rate Equations**

The rate of a chemical reaction depends on the frequency and force of collisions between molecules. Changes in the concentration and the temperature of the system are observed experimentally to establish a rate equation for the reaction.

(1) For the hypothetical reaction:

$$2A + 4B \Rightarrow 5C + 3D$$

$$\text{Rate} = -\frac{1}{2}\frac{\Delta[A]}{\Delta t} = -\frac{1}{4}\frac{\Delta[B]}{\Delta t} = +\frac{1}{5}\frac{\Delta[C]}{\Delta t} = +\frac{1}{3}\frac{\Delta[D]}{\Delta t}$$

Note: Δ represents 'change in'

Two (2) A's are reacting, and A is disappearing as time goes on. The *minus* sign is used in the rate equation to show A is being consumed. One-half ($\frac{1}{2}$) of the rate of disappearance (*minus* sign) of A with time will give the specific rate of reaction. One-quarter ($\frac{1}{4}$) the rate of disappearance (*minus* sign) of B, one-fifth ($\frac{1}{5}$) of the rate of appearance (*plus* sign) of the product C and one-third ($\frac{1}{3}$) of the rate of appearance (*plus* sign) of the product D will each give the same rate for the reaction.

(2) For the hypothetical reaction:

$$3A + B \Rightarrow 2C + D$$

$$\text{Rate} = -\frac{1}{3}\frac{\Delta[A]}{\Delta t} = -\frac{1}{1}\frac{\Delta[B]}{\Delta t} = +\frac{1}{2}\frac{\Delta[C]}{\Delta t} = k[A]^m[B]^n$$

In the sample reaction:
m is the order with respect to A
n is the order with respect to B
m + n is the overall order

The **reaction order** is determined experimentally. In this case it is equal to m + n. It is an exponent of the reactant molarity. The values of m and n may or may not be the same as the stoichiometric coefficients of the balanced equation. Often they differ. The values of m and n are frequently small integers such as 1, 2 or 3. Orders of $\frac{1}{2}$ (square root) and –1 (inversely proportional) are not uncommon.

The **specific rate constant**, k, is dependent on reaction temperature, and must be determined using experimental data.

Examples

1. Given the reaction: $N_2O_3(g) \Rightarrow NO(g) + NO_2(g)$

    Note: The abbreviation for seconds is 's'.

Data	Pressure (N_2O_3)	Rate
Run 1	0.0015 atm	0.00917 atm s^{-1}
Run 2	0.0020 atm	0.0122 atm s^{-1}
Run 3	0.0025 atm	0.0153 atm s^{-1}
Run 4	0.0030 atm	0.0183 atm s^{-1}

    (a) The rate equation is determined by observing the effect of a change in the reactant concentration of the reaction rate. Using Runs 1 and 4 makes things simpler since there is a *doubling* of concentration because of the doubling of the partial pressure—partial pressure is a measure of concentration.

    $$\left(\frac{P_{N_2O_3} \text{ (Run4)}}{P_{N_2O_3} \text{ (Run1)}}\right)^m = \left(\frac{\text{Rate (Run4)}}{\text{Rate (Run1)}}\right)$$

    $$\left(\frac{0.0030}{0.0015}\right)^m = \left(\frac{0.0183}{0.00917}\right)$$

    $$(2.0)^m = (2.0)$$

    The doubling of the partial pressure has doubled the rate. The power of the partial pressure is one (m=1). The reaction is 'first order'.

    (b) After the order of the reactants has been determined, it is possible to determine the value of the specific rate constant, k, including its units.

    Using the experimental data of Run 1:
    Rate = k $[P_{N_2O_3}]^1$
    0.00917 atm s^{-1} = k × 0.0015 atm
    k = 6.11 s^{-1}

2. Given the hypothetical reaction:
$$A + 2B + C \Rightarrow Products$$

Data	[A]	[B]	[C]	Time
Run 1	0.20 M	0.20 M	0.20 M	160 s
Run 2	0.20 M	0.40 M	0.10 M	160 s
Run 3	0.40 M	0.40 M	0.10 M	40 s
Run 4	0.40 M	0.40 M	0.20 M	20 s

(a). In getting ready to write the rate equation, it should be noted that *time* rather than rate is given in this case. The time is a reciprocal in reaction rate (atm s^{-1}, moles L s^{-1}, etc). Although the actual rate could be calculated, it is more convenient and just as accurate to use the reciprocal of time as the reaction rate when getting the order of the reaction.

To get the order of **A** use Run 2 and Run 3. In these runs [A] is changing but both [B] and [C] are not changing.

$$\left(\frac{[A](Run\ 3)}{[A](Run\ 2)}\right)^m = \left(\frac{\frac{1}{Time}(Run\ 3)}{\frac{1}{Time}(Run\ 2)}\right)$$

$$\left(\frac{0.40}{0.20}\right)^m = \left(\frac{\frac{1}{40}}{\frac{1}{160}}\right) = \left(\frac{160}{40}\right)$$

$$(2.0)^m = 4.0$$

The order of A is 2.

To get the order of **C** use Run 3 and Run 4. In these runs [C] is changing but both [A] and [B] are not changing.

$$\left(\frac{[C](Run\ 4)}{[C](Run\ 3)}\right)^n = \left(\frac{\frac{1}{Time}(Run\ 4)}{\frac{1}{Time}(Run\ 3)}\right)$$

$$\left(\frac{0.20}{0.10}\right)^n = \left(\frac{\frac{1}{20}}{\frac{1}{40}}\right) = \left(\frac{40}{20}\right)$$

$$(2.0)^n = 2.0$$

The order of C is 1.

To get the order of **B** use Run 1 and Run 2. In these runs [B] and the [C] are both changing. [A] is not changing and the order of C is already know.

$$\left(\frac{[B](\text{Run 2})}{[B](\text{Run 1})}\right)^r \left(\frac{[C](\text{Run 2})}{[C](\text{Run 1})}\right)^1 = \left(\frac{\frac{1}{\text{Time}}(\text{Run 2})}{\frac{1}{\text{Time}}(\text{Run 1})}\right)$$

$$\left(\frac{0.40}{0.20}\right)^r \left(\frac{0.10}{0.20}\right)^1 = \left(\frac{\frac{1}{160}}{\frac{1}{160}}\right) = \left(\frac{160}{160}\right)$$

$$(2.0)^r \left(\frac{1.0}{2.0}\right)^1 = 1.0$$

The order of B is 1.

The overall order is 4, and the rate equation is:

$$\text{Rate} = k[A]^2[B]^1[C]^1$$

(b) The value of the rate constant, k, can be determined by using Run 1 and the rate of disappearance of A.

The unit of k is the reciprocal of time and molarity (time^{-1}; molarity^{-1}). The exponent of the molarity is *one less* than the overall order of 4.

$$\frac{0.20 \text{ M}}{160 \text{ s}} = k[0.20 \text{ M}]^2[0.20 \text{ M}]^1[0.20 \text{ M}]^1$$

$$k = 0.78 \text{ L}^3 \text{ mol}^{-3} \text{ s}^{-1}$$

2. First Order Reactions

The half-life relationship used in Chapter 1 for nuclear reactions is the same as that of a first-order chemical reaction. Whenever the rate of a chemical reaction is found to be first-order, the half-life equations will apply.

$$\ln \frac{[N]}{[N_o]} = -kt$$

$$k = \frac{0.693}{t_{1/2}}$$

The half-life is the time required for half of the reactant to be consumed. It is *independent* of the initial concentration of the reactant.

Example

The decomposition of hydrogen peroxide, H_2O_2, at 70°C is first-order and has a rate constant of 0.0347 min^{-1}

$$2H_2O_2(aq) \Rightarrow 2H_2O(l) + O_2(g)$$

How long will it take for 70.0% of a sample of hydrogen peroxide to decompose (30.0% remains), and what is the half-life of H_2O_2 at 70°C?

$$\ln \frac{[N]}{[N_o]} = -kt$$

$\ln 0.300 = -0.0347 \text{ min}^{-1} \times t$

Answer: t = 35 minutes

$$t_{1/2} = \frac{0.693}{k}$$

$$t_{1/2} = \frac{0.693}{0.0347 \text{ min}^{-1}}$$

Answer: $t_{1/2}$ = 20 minutes

3. Activation Energy and Catalysis

The collision theory of reaction rates describes reactions in terms of collisions between reacting molecules. To be effective, a collision must be both aligned properly and energetic enough to form an intermediate species called the 'activated complex'. The number of effective collisions is a small fraction of the total number of collisions between reacting molecules. Most of the molecules bounce off one another and remain unchanged.

The 'activation energy', E_a, is the minimum energy required to cause a reaction between molecules colliding with the proper geometry.

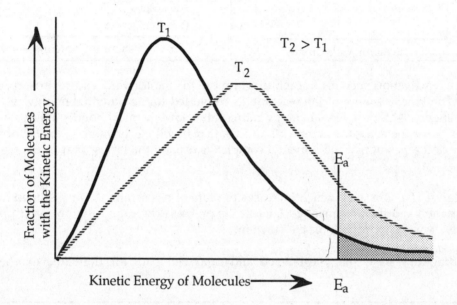

The kinetic energy diagram, sometimes called the Maxwell-Boltzmann distribution, illustrates a reason why it is often stated that a small 10°C increase in temperature will *double* the rate of reaction. Although there is an increase in molecular collisions because of an increase in temperature (kinetic energy), this fact alone *cannot* account for the dramatic increase in reaction rate. The explanation lies in the fact the number of molecules that will have *effective collisions*—with an energy equal to or greater than the activation energy, E_a—has more than doubled.

A potential energy diagram shows how the energy of the reactants, activated complex, and products are related for an exothermic reaction and for an endothermic reaction.

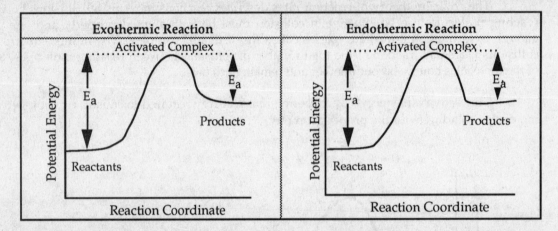

The two activation energies in each diagram are for the forward and reverse reaction, respectively. The kinetic energy of the reactants is converted to the potential energy required as activation energy. A high potential energy represents a low chemical bond stability. A high potential energy for the activated complex indicates that it will easily come apart to form the products. The decomposition of the activated complex is usually the fastest part of the reaction mechanism.

A catalyst is a substance which increases the rate of a reaction. It is consumed in one step of the reaction and then regenerated later in the process. The catalyst is not used up, but provides a new, lower energy path for the reaction.

The diagrams show the uncatalyzed and catalyzed path for the same exothermic reaction.

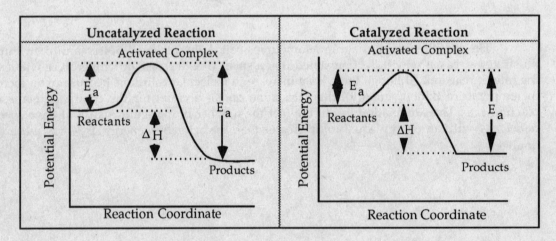

The catalyzed reaction has a lower activation energy, and there are significantly more molecules with the kinetic energy required to form the activated complex and then product.

An additional point illustrated by this figure is the *same* heat of reaction (ΔH) for both the uncatalyzed and catalyzed reactions. This is because the energy of the reactants and the products are unchanged since it is the same reaction.

Reaction Kinetics

4. Heat Effects and the Arrhenius Equation

The Arrhenius equation, first shown to be valid in 1889 by the Swedish chemist Svante Arrhenius, shows the relationship between the rate constant, k, and temperature, T.

$$k = Ae^{-\frac{E_a}{RT}}$$

where: A is a constant
         E_a is the activation energy

The plot of ln k vs $\frac{1}{T}$ is a straight line. The slope of this line is equal to $-\frac{E_a}{R}$. Many AP Chemistry classes have done a laboratory experiment—the iodation or bromination of acetone—that allows the estimation of the activation energy of a reaction. Some texts develop this equation using logs, but modern calculators have both 'log' and natural log (ln x) functions. The use of the Arrhenius equation in the base 'e' form precludes the need to use the conversion between the two log forms (ln x = 2.303 log x).

An integrated version of the Arrhenius equation is used to determine:

(a) The value of the rate constant at different temperatures when the activation energy is known.

(b) The value of the activation energy when the rate constants at two temperatures are known.

Remember the form of this equation. It is also used in Chapter 6 to calculate the change with temperature of the equilibrium constant, K_p, by using the value for the heat of reaction, ΔH, instead of E_a.

$$\ln\frac{k_2}{k_1} = -\frac{E_a}{R}\left[\frac{1}{T_2} - \frac{1}{T_1}\right]$$

The value of the gas constant, R, used to calculate the activation energy is 8.314 J mol^{-1} K^{-1}. Both rate constants must have the same dimensions.

Example:

(a) Calculate the activation energy in kJ for the reaction where:

k_1 is 5.00×10^{-2} mol^{-1} s^{-1} at 287°C

k_2 is 6.80×10^{0} mol^{-1} s^{-1} at 333°C

To calculate the activation energy, first write the Arrhenius equation.

$$\ln \frac{k_2}{k_1} = -\frac{E_a}{R}\left[\frac{1}{T_2} - \frac{1}{T_1}\right]$$

Then:

$$\ln \frac{6.80 \times 10^{0}}{5.00 \times 10^{-2}} = -\frac{E_a}{8.314 \text{ J mol}^{-1} \text{ K}^{-1}}\left[\frac{1}{606} - \frac{1}{560}\right]$$

$$4.91 = -\frac{E_a}{8.314 \text{ J mol}^{-1} \text{ K}^{-1}}\left[1.65 \times 10^{-3} - 1.79 \times 10^{-3}\right]$$

$E_a = +301{,}000$ J $= 301$ kJ

(b) What will be the value of the rate constant at 313°C?

$$\ln \frac{k_2}{5.00 \times 10^{-2}} = -\frac{301{,}000 \text{ J}}{8.314 \text{ J mol}^{-1} \text{ K}^{-1}}\left[\frac{1}{586} - \frac{1}{560}\right]$$

$$\ln \frac{k_2}{5.00 \times 10^{-2}} = 2.87$$

Be sure during the review that a calculator is used to practice each step of this calculation, and similar practice solutions. The calculator is used at this point to find a value which is the inverse ln x of 2.87 (or, on some calculators, e^x of 2.87).

$$\frac{k_2}{5.00 \times 10^{-2}} = 17.6$$

$k_2 = 17.6 \times 5.00 \times 10^{-2}$

$k_2 = 8.80 \times 10^{-1}$ mol^{-1} s^{-1} at 313°C

Reaction Kinetics

5. Reaction Mechanism

A series of steps (usually 2 to 4) that add up to the overall reaction is the reaction mechanism.

1. OVERALL REACTION: $2 NO_2 + F_2 \Rightarrow 2 NO_2F$
   Mechanism:
   1. $NO_2 + F_2 \Rightarrow NO_2F + F$ (slow)
   2. $F + NO_2 \Rightarrow NO_2F$ (fast)

   PRINCIPLES INVOLVED:
   a. The *slow step* is the rate-determining step of the overall reaction, and determines the rate law for the overall reaction.
   b. An *intermediate* is formed from the reactants and is involved in the process, but does not appear in the overall reaction. Fluorine, F, is an intermediate in this mechanism.

   RATE EQUATION: Rate = $k_1[NO_2][F_2]$

2. OVERALL REACTION: $2 O_3 \Rightarrow 3 O_2$

   RATE EQUATION: Rate = $k \dfrac{[O_3]^2}{[O_2]}$ where: $k = K_1 \times k_2$

   MECHANISM:
   1. $O_3 \Leftrightarrow O_2 + O$ (fast equilibrium)
   2. $O_3 + O \Rightarrow 2 O_2$ (slow)

   PRINCIPLE INVOLVED:
   Use the equilibrium step to *remove intermediates* from the rate equation. It is the involvement of intermediates that makes it possible to propose these mechanisms, but an intermediate *should not* appear in the rate law equation.

   A. Rate = $k_2[O_3][O]$ (from the slow step)
   B. $K_1 = \dfrac{[O_2][O]}{[O_3]}$ (from the equilibrium step)

   $[O] = K_1 \dfrac{[O_3]}{[O_2]}$

   C. Rate = $K_1 k_2 \dfrac{[O_3]^2}{[O_2]}$ (combine A and B)

Examples

Given the overall reaction and the rate equation:
(a) Propose a three-step mechanism.
(b) Use the mechanism to derive a rate equation in terms of reactant concentrations only. Determine how the overall rate constant, k, relates to the rate constants and equilibrium constants of the mechanism steps.

1. OVERALL REACTION: $I^- + OCl^- \Rightarrow OI^- + Cl^-$ in basic solution

   RATE EQUATION: $\text{Rate} = k \dfrac{[I^-][OCl^-]}{[OH^-]}$

   Answer:
   MECHANISM:
   1. $OCl^- + H_2O \Leftrightarrow HOCl + OH^-$ (fast equilibrium)
   2. $I^- + HOCl \Rightarrow HOI + Cl^-$ (slow)
   3. $OH^- + HOI \Rightarrow H_2O + OI^-$ (fast)

   A. $\text{Rate} = k_2[I^-][HOCl]$ (the slow reaction is step 2)

   B. $K_1 = \dfrac{[HOCl][OH^-]}{[OCl^-]}$ $[HOCl] = K_1 \dfrac{[OCl^-]}{[OH^-]}$ (from the equilibrium)

   C. $\text{Rate} = k \dfrac{[I^-][OCl^-]}{[OH^-]}$ $k = K_1 k_2$ (combing A and B)

2. OVERALL REACTION: $3HNO_2 \Rightarrow H^+ + NO_3^- + 2NO + H_2O(l)$

   RATE EQUATION: $\text{Rate} = k \dfrac{[HNO_2]^4}{[NO]^2}$

   Answer:
   MECHANISM:
   1. $4HNO_2 \Leftrightarrow 2NO_2 + 2NO + 2H_2O(l)$ (fast, equilibrium)
   2. $2NO_2 \Leftrightarrow N_2O_4$ (fast, equilibrium)
   3. $N_2O_4 + H_2O(l) \Rightarrow H^+ + NO_3^- + HNO_2$ (slow)

   A. $\text{Rate} = k_3[N_2O_4]$ (the slow reaction is step 3)

   B. Liquid water, $H_2O(l)$, is not included in the equilibrium equation for step 2.

   $K_2 = \dfrac{[N_2O_4]}{[NO_2]^2}$ $[N_2O_4] = K_2[NO_2]^2$ (N_2O_4 is an intermediate)

   C. $K_1 = \dfrac{[NO_2]^2[NO]^2}{[HNO_2]^4}$ $[NO_2]^2 = K_1 \dfrac{[HNO_2]^4}{[NO]^2}$ (NO_2 is an intermediate)

   D. $\text{Rate} = k_3[N_2O_4] = k_3 K_2[NO_2]^2 = k_3 K_2 K_1 \dfrac{[HNO_2]^4}{[NO]^2}$ (combine A, B and C)

   $k = K_1 K_2 k_3$

Chapter 6
Equilibrium

1. Equilibrium Constants

The equilibrium constant, K, expresses arithmetically the extent to which a reaction will proceed at a given temperature. The equilibrium law is defined in terms of the concentrations or, in the case of gases, the pressures that exist at equilibrium.

The equilibrium law equation takes two forms for the gaseous equilibrium reaction:
$$2CO_2(g) \Leftrightarrow 2CO(g) + O_2(g)$$

(a) When concentrations are expressed as molarity:
$$K_c = \frac{[CO]^2[O_2]}{[CO_2]^2}$$

(b) When concentrations are expressed as partial pressures:
$$K_p = \frac{P_{CO}^2 \, P_{O_2}}{P_{CO_2}^2}$$

The equations relating K_c and K_p are:
$$K_c = K_p \left[\frac{1}{RT}\right]^{\Delta n} \qquad K_p = K_c [RT]^{\Delta n}$$

where Δn = (moles of gaseous product) − (moles of gaseous reactant)

Example
A 3:1 starting mixture of hydrogen, H_2, and nitrogen, N_2, comes to equilibrium at 450.°C. The mixture at equilibrium is 9.6% NH_3, 22.6% N_2, and 67.8% H_2 by volume and at 60.0 atmospheres.
What is the value of K_p and K_c for this reaction?
$$N_2(g) + 3H_2(g) \Leftrightarrow 2NH_3(g)$$

Dalton's law of partial pressures is used to calculate the equilibrium partial pressures.

NH_3:	0.096 × 60.0	=	5.8 atm
N_2:	0.226 × 60.0	=	13.5 atm
H_2:	0.678 × 60.0	=	40.7 atm

$$K_p = \frac{P_{NH_3}^2}{P_{N_2} P_{H_2}^3} = \frac{(5.8)^2}{(13.5)(40.7)^3} = 3.7 \times 10^{-5}$$

$$K_c = K_p \left[\frac{1}{RT}\right]^{\Delta n} = 3.7 \times 10^{-5} \left[\frac{1}{0.0821 \times 723}\right]^{-2} = 0.13$$

2. Concentration Changes in Equilibrium Reactions

If the concentration (pressure) of one of the constituents changes at constant temperature, the equilibrium will not be reestablished until all of the equilibrium concentrations change, and the ratio:

$$\frac{[\text{PRODUCTS}]}{[\text{REACTANTS}]} \text{ is equal to K.}$$

To keep track of the changes occurring as equilibrium is established or re-established, the basic thesis of the 'mole link' procedure used in Chapter 3, 'Stoichiometry', must be modified. It must be taken into account that at equilibrium none of the species is at zero (0) concentration. All of the constituents of the reaction are present to the extent allowed by the numerical value of K. The concentration change is followed by using a table that shows the:

> START (starting concentrations)
> Δ (delta; the change in concentrations)
> FINISH (equilibrium concentrations)

Some problems will seem to require that the complete solution will involve the use of the quadratic equation. When the equilibrium constant, K, is less than 0.01 (K < 0.01) or greater than 100 (K > 100) an important approximation can be made for the dilute solutions encountered in most equilibrium problems. Unknown values that are added to or subtracted from known values are comparatively small and are ignored.

Examples

1. Determine the starting pressures for the equilibrium problem in the last section. The percent composition at equilibrium is 9.6% NH_3, 22.6% N_2, and 67.8% H_2 by volume. The temperature is 450.°C and the total pressure is 60.0 atmospheres.

   Dalton's law of partial pressures predicts the equilibrium pressures will be:

NH_3:	0.096×60.0	=	5.8 atm
N_2:	0.226×60.0	=	13.5 atm
H_2:	0.678×60.0	=	40.7 atm

   A key to the solution is to note:

   a. There is no product, NH_3, at the start of the reaction.

   b. The change (5.8 atm) of the NH_3 all occurs as equilibrium is established.

   c. During the reaction the stoichiometry of the reaction is followed, and 'mole links' are used extensively.

	$N_2(g)$	$+ 3H_2(g)$	$\Leftrightarrow$	$2NH_3(g)$
Start	16.4 atm	49.4 atm		0.0 atm
Δ	− 2.9	− 8.7		+ 5.8
Finish	13.5	40.7		5.8

2. The reaction $H_2(g) + I_2(g) \Leftrightarrow 2HI(g)$ has an equilibrium constant of 54.4 at 355°C. At 355°C and a total pressure of 0.50 atm, 0.20 mole each of H_2 and I_2 are converted to HI.

a. What percent of the iodine is converted to product?

	$H_2(g)$	$+ I_2(g)$	$\Leftrightarrow$ 2HI(g)
Start	0.25 atm	0.25 atm	0.0 atm
Δ	–x	–x	+2x
Finish	0.25 – x	0.25 – x	2x

$$K_p = \frac{P_{HI}^2}{P_{H_2} P_{I_2}} = \frac{(2x)^2}{(0.25-x)^2} = 54.4$$

x = 0.20 atm

% conversion = (0.20 ÷ 0.25) × 100 = **80.%**

b. What is the numerical value of K_c?

Since $\Delta n = 0$ for this reaction, $K_c = K_p =$ **54.4**

3. The dissociation of H_2S at 900. K has a K_p of 0.0100 atm. What is the percent dissociation at this temperature?

Let x = dissociated hydrogen sulfide

	$2H_2S(g)$	$\Leftrightarrow$	$2H_2(g)$	$+ S_2(g)$
Start	100. atm		0.0 atm	0.0 atm
Δ	–x		+x	+x/2
Finish	100. – x		x	x/2

The x subtracted from 100, (100 – x), is ignored.

$$K_p = 0.0100 = \frac{P(Products)}{P(Reactants)} = \frac{(x)^2(x/2)}{(100.)^2}$$

x = % dissociation = **5.84%**

The answer would have been 5.6% if x was not ignored.

3. LeChatelier's Principle

The effect of changes in pressure, concentration, and temperature on the constituent species and on the equilibrium constant can be predicted. LeChatelier's Principle states that a system in equilibrium will react to a stress in a way that relieves the stress.

CONCENTRATION
If the concentration of a chemical on the reactant side of the reaction is increased, then the reaction to form products is favored and proceeds at a higher reaction rate. The concentration of all the reactants is reduced, relieving the stress. This will result in a 'shift' toward increased concentration of the products.

PRESSURE
If the pressure of a system containing gaseous molecules is increased, then the number of molecules must be reduced to relieve the stress. The reaction will 'shift' to the side with the smallest moles of gaseous molecules.

TEMPERATURE
The reaction to products is favored because of a higher forward reaction rate when heat is added to an endothermic reaction (heat is on the reactant side of the reaction). The heat is 'used up' by the endothermic reaction, relieving the stress. The reaction rate is equal after a 'shift' toward higher concentrations of products and lower concentrations of reactants. The equilibrium constant, K, is larger when heat is added to an endothermic reaction.

When heat is added to an exothermic reaction (heat is on the product side of the reaction), the reaction rate of the products is 'favored'. The result is a decreased product concentration and an increased reactant concentration. The equilibrium constant, K, is smaller.

CATALYSIS
Catalysts *do not* affect equilibrium concentrations. The addition of the catalyst speeds up the rate of both the forward and reverse reactions. At equilibrium these reaction rates are already equal. The effect of the catalyst is to make these equal rates faster.

Example
Refer to the three gaseous equilibrium reactions shown. All species are gaseous, (g).

Reaction A:	N_2O_4	$\Leftrightarrow 2NO_2$	ΔH is positive
Reaction B:	$CO_2 + H_2$	$\Leftrightarrow CO + H_2O$	ΔH is positive
Reaction C:	$SO_2 + \frac{1}{2}O_2$	$\Leftrightarrow SO_3$	ΔH is negative

Fill in the chart by stating the direction of the shift and the effect on the numerical value of K for each change listed in the table. The effects on the reaction mixture occur after it has reached its initial equilibrium condition.

The symbols used in the table are:

'+', '−', and 'none' representing INCREASE, DECREASE, and NO EFFECT, respectively.

'0', 'R', and 'L' representing NO EFFECT, SHIFT RIGHT, and SHIFT LEFT respectively.

	Reaction A		Reaction B		Reaction C	
	Effect on:		Effect on:		Effect on:	
	Position	K_p	Position	K_p	Position	K_p
Increase in pressure	Left	0	None	0	Right	0
Increase in temperature	Right	+	Right	+	Left	−
Increase in the partial pressure of XO_2 gas	Left	0	Right	0	Right	0
A catalyst is added	None	0	None	0	None	0
Helium is added; the total pressure is the same.	None	0	None	0	None	0

4. **Temperature Change and the van't Hoff Equation**

The numerical value of the equilibrium constant, K, depends only on the temperature. An increase or a decrease in K's numerical value is predicted by LeChatelier's Principle for either an endothermic reaction (K increases with temperature) or an exothermic reaction (K decreases with temperature).

The **van't Hoff equation** has a form similar to the Arrhenius equation used in Chapter 5—except that heat of reaction, ΔH, is substituted for activation energy, E_a. The van't Hoff equation is used to find the new value of K when the temperature is changed.

$$\ln \frac{K_2}{K_1} = -\frac{\Delta H^\circ}{R}\left[\frac{1}{T_2} - \frac{1}{T_1}\right]$$

A form of the van't Hoff equation, called the **Clausius Clapeyron** equation is used to find the vapor pressure of a substance at a different temperature or to find the heat of vaporization, $\Delta H_{vaporization}$. This equation could be used to solve a State of matter (vapor pressure), an Equilibrium (vapor-liquid equilibrium), or a Thermodynamic (enthalpy of vaporization, ΔH) problem. The sample problem shown in this chapter shows the similarity in the form of the van't Hoff and Clausius-Calpeyron equations.

$$\ln \frac{P_2^o}{P_1^o} = -\frac{\Delta H^o_{vap}}{R}\left[\frac{1}{T_2} - \frac{1}{T_1}\right]$$

Examples

1. The reaction:

$$PCl_3(g) + Cl_2(g) \Leftrightarrow PCl_5(g)$$

is exothermic with a standard heat of reaction, $\Delta H°$, of -92.9 kJ mol^{-1}. The value of the equilibrium constant, K, is 0.562 atm^{-1} at 250.°C. What is its value at 200.°C?

$$\ln \frac{K_2}{K_1} = -\frac{\Delta H°}{R}\left[\frac{1}{T_2} - \frac{1}{T_1}\right]$$

$$\ln \frac{K_2}{0.562} = -\frac{-92,900}{8.314}\left[\frac{1}{473} - \frac{1}{523}\right] = 2.26$$

$$K_2 = 0.562 \times 9.57$$

$$K_2 = \mathbf{5.38}$$

The equilibrium constant for this exothermic reaction is higher at the lower temperature. An increase in temperature favors an endothermic reaction—in this case the endothermic reaction is the reverse reaction.

2. The vapor pressure of water at 25°C is 23.8 mmHg, and the boiling point of water is 100.°C.
   (a) Estimate the standard heat of vaporization, $\Delta H°_{vap}$, of water.
   (b) Estimate the vapor pressure of water at 50°C.

(a) $$\ln \frac{P_2°}{P_1°} = -\frac{\Delta H°_{vap}}{R}\left[\frac{1}{T_2} - \frac{1}{T_1}\right]$$

$$\ln \frac{23.8}{760} = -3.46 = -\frac{\Delta H°_{vap}}{8.314}\left[\frac{1}{298} - \frac{1}{373}\right] = -\frac{\Delta H°_{vap}}{8.314}\left[6.75 \times 10^{-4}\right]$$

$$\Delta H_{vap} = \frac{(-3.46 \times -8.314)}{6.75 \times 10^{-4}} = 42,600 \text{ J} = \mathbf{42.6 \text{ kJ}}$$

(b) $$\ln \frac{P_2}{760} = -\frac{+42,600}{8.314}\left[\frac{1}{323} - \frac{1}{373}\right] = -2.13$$

$$\frac{P_2}{760} = 0.119$$

$$P_2 = \mathbf{90.6 \text{ mmHg}}$$

Equilibrium 61

5. Solubility Equilibria and K_{sp}

When a 'slightly soluble' or 'insoluble' salt is mixed with water, a saturated solution results and solubility equilibrium is established. The rate of dissociation of the ions from the solid equals the rate of precipitation of the salt.

The equilibrium rules apply to the dissolving of slightly soluble salts to form saturated solutions.

$$\text{Salt(s)} \Rightarrow m \text{ Cation(aq)} + n \text{ Anion(aq)}$$

The unique equilibrium constant, solubility product (K_{sp}) is used to describe this common equilibrium condition. The equation for K_{sp} takes the form:

$$K_{sp} = [\text{cation}]^m[\text{anion}]^n$$

The concentration of all solids is constant and does not appear in the equilibrium law equation. The amount of solid in a saturated solution is not important.

'X' is generally used to denote the molar solubility of a slightly soluble salt. Two commonly found types of problems appear frequently and will be illustrated in the following examples. The 'x^2–type' of dissociation is applicable to those salts which break up into two ions (e.g. AgCl, $BaSO_4$, and $PbCO_3$). Salts which break up into three ions are of the '$4x^3$–type' (e.g. Ag_2CrO_4, CaF_2, and $Zn(OH)_2$). Recognition of a 'type' will result in a quicker response to solubility questions.

Examples

1. The solubility of barium chromate, $BaCrO_4$, in water is 1.3×10^{-5} M at 25°C. What is the value of K_{sp} at this temperature?

	$BaCrO_4(s)$	$\Leftrightarrow$	$Ba^{2+}(aq)$	$+ CrO_4^{2-}(aq)$
Start	Some		0.0 M	0.0 M
Δ	$-x$		$+x$	$+x$
Finish	Some $- x$		x	x

The molar solubility, x, is 1.3×10^{-5} M

$K_{sp} = [Ba^{2+}][CrO_4^{2-}] = [x][x] = x^2$

$K_{sp} = [1.3 \times 10^{-5}]^2 = 1.7 \times 10^{-10}$

2. The solubility product of silver chromate, Ag_2CrO_4, in water is 1.2×10^{-12} at 15°C. What is the molar solubility at this temperature?

	$Ag_2CrO_4(s)$	$\Leftrightarrow$	$2Ag^+(aq)$	$+ CrO_4^{2-}(aq)$
Start	Some		0.0 M	0.0 M
Δ	$-x$		$+2x$	$+x$
Finish	Some $- x$		$2x$	x

$K_{sp} = [Ag^+]^2[CrO_4^{2-}] = [2x]^2[x] = 4x^3 = 1.2 \times 10^{-12}$
x, the *molar solubility* = 6.7×10^{-5} M

6. Common Ion Effect

It is possible to shift the solubility equilibrium to favor the reactants. This is accomplished by adding a solution which contains an ion in common—the 'common ion'—with the salt. The result will be a saturated solution which has:
    a. a lower solubility for the starting solid
    b. more undissolved solid and a lower ion concentration of the other ion(s)

Examples
What is the solubility of MgF_2?
For MgF_2, the $K_{sp} = 6.4 \times 10^{-9}$ at 27°C.

(a) **In pure water.**

	$MgF_2(s)$	$\Leftrightarrow$	$Mg^{2+}(aq)$	$+ 2F^-(aq)$
Start	Some		0.0 M	0.0 M
Δ	$-x$		$+x$	$+2x$
Finish	Some $- x$		x	$2x$

$K_{sp} = [Mg^{2+}][F^-]^2 = [x][2x]^2 = 4x^3 = 6.4 \times 10^{-9}$
x, the *molar solubility*, = 1.2×10^{-3} M

(b) **In a 0.10 M solution of NaF** (all sodium salts are soluble in water).

	$MgF_2(s)$	$\Leftrightarrow$	$Mg^{2+}(aq)$	$+ 2F^-(aq)$
Start	Some		0.0 M	0.10 M
Δ	$-x$		$+x$	$+2x$
Finish	Some $- x$		x	0.10
Note: 2x is small compared to 0.10, and is ignored.				

$K_{sp} = [Mg^{2+}][F^-]^2 = [x][0.10]^2 = 6.4 \times 10^{-9}$
x, the *molar solubility*, = 6.4×10^{-7} M
This is **0.05%** of the solubility of MgF_2 in pure water.

(c) In a 0.10 M solution of $Mg(NO_3)_2$.

	$MgF_2(s)$	$\Leftrightarrow$	$Mg^{2+}(aq)$	$+ 2F^-(aq)$
Start	Some		0.10 M	0.0 M
Δ	$-x$		$+x$	$+2x$
Finish	Some $-x$		0.10	$2x$
Note: x is small compared to 0.10, and is ignored.				

$K_{sp} = [Mg^{2+}][F^-]^2 = [0.10][2x]^2 = 6.4 \times 10^{-9}$
x, the *molar solubility*, $= 1.3 \times 10^{-4}$ M
This is **10%** of the solubility of MgF_2 in pure water.

7. **Selective Precipitation**

The difference in the solubility of two salts containing a common cation (positive ion) can be used to separate a pair of anions (negative ions). Similarly, cations can be separated by adding a common anion.

Example
Water soluble silver nitrate, $AgNO_3(aq)$, is added to a solution which is 0.10 M in sodium chloride, NaCl, and 0.010 M in potassium chromate, K_2CrO_4. Assume no dilution caused by the addition of $AgNO_3$.

The K_{sp} of AgCl is 1.6×10^{-10} at 25°C.
The K_{sp} of Ag_2CrO_4 is 9.0×10^{-12} at 25°C.
(a) Which precipitates first, AgCl or Ag_2CrO_4?
(b) What will be the molarity of silver ion, $[Ag^+]$, when precipitation begins?
(c) What is the molarity of the chloride ion, $[Cl^-]$, when the Ag_2CrO_4 first starts to precipitate?

(a) The molarity of the silver ion, $[Ag^+]$, is calculated for each of the two equilibrium reactions. The precipitation which occurs first is that of the salt with the smaller equilibrium concentration of Ag^+.

The solubility equilibrium reaction of silver chloride, AgCl, shows for these data:

	$AgCl(s)$	$\Leftrightarrow$	$Ag^+(aq)$	$+ Cl^-(aq)$
Start	Some		0.0 M	0.10 M
Δ	$-x$		$+x$	$+x$
Finish	Some $-x$		x	0.10
Note: x is small compared to 0.10, and is ignored.				

$K_{sp} = [Ag^+][Cl^-] = [x][0.10] = 1.6 \times 10^{-10}$
x, the $[Ag^+] = \mathbf{1.6 \times 10^{-9}}$ **M**

The silver chromate, Ag_2CrO_4, equilibrium reaction shows for these data:

	$Ag_2CrO_4(s)$ ⇔	$2Ag^+(aq)$	$+ CrO_4^{2-}(aq)$
Start	Some	0.0 M	0.010 M
Δ	$-x$	$+2x$	$+x$
Finish	Some $- x$	$2x$	0.010
Note: x is small compared to 0.010, and is ignored.			

$K_{sp} = [Ag^+]^2[CrO_4^{2-}] = [2x]^2[0.010] = 9.0 \times 10^{-12}$
$2x = $ the $[Ag^+] = 3.0 \times 10^{-5}$ M

(b) The silver chloride, AgCl, starts precipitating out first, at a molarity of silver ion, $[Ag^+]$, equal to 1.6×10^{-9} M

(c) Part (a) shows that in order to precipitate Ag_2CrO_4 the $[Ag^+]$ must be 3.0×10^{-5} M.

$K_{sp} = [Ag^+][Cl^-] = [3.0 \times 10^{-5}][Cl^-] = 1.6 \times 10^{-10}$
The $[Cl^-] = 5.3 \times 10^{-6}$ M

This is 0.005% of the starting chloride ion molarity of 0.10. The separation of Cl^- by precipitation as the silver salt is effective.

8. **Mixtures of Two Solutions and 'Bounce Back'**

The mixture of two solutions may or may not result in a precipitate. The outcome of such a mixture can be predicted by using the solubility product constant of the solid.

If the product $[cation]^m[anion]^n$ is:

    a. greater than (>) K_{sp}, the solution *exceeds saturation*. Precipitation of the solid *will* occur until the product of the ion concentrations is equal to K_{sp}.

    b. equal to (=) K_{sp}, the solution is *saturated*.

    c. less than (<) K_{sp}, the solution is *unsaturated*. Precipitation *will not* occur until the product of the ion concentrations exceeds the K_{sp}.

Equilibrium

Example

1. A solution with a final volume of 20.0 mL is made by mixing 10.0 mL of 0.10 M $Pb(NO_3)_2$ and 10.0 mL of 0.0010 M Na_2SO_4. The K_{sp} of $PbSO_4$ is 1.06×10^{-8}.

$$PbSO_4(s) \Leftrightarrow Pb^{2+}(aq) + SO_4^{2-}(aq)$$

Will a precipitate form?

$[Pb^{2+}] = \dfrac{10.0 \text{ mL} \times 0.10 \text{ M}}{20.0 \text{ mL}} = 0.0500 \text{ M}$

$[SO_4^{2-}] = \dfrac{10.0 \text{ mL} \times 0.0010 \text{ M}}{20.0 \text{ mL}} = 0.000500 \text{ M}$

$[Pb^{2+}][SO_4^{2-}] = [0.0500][0.00050] = \mathbf{2.50 \times 10^{-5}}$

$K_{sp} = 1.06 \times 10^{-8}$

The value $\mathbf{2.50 \times 10^{-5}}$ for $[Pb^{2+}] \times [SO_4^{2-}]$ is greater than (>) K_{sp}. Precipitation *will* occur.

Precipitation reaction calculations make use of the solubility equilibrium expression to find the concentration of ions in the final solution. When two solutions whose product of concentration ($[cation]^m \times [anion]^n$) exceeds K_{sp} are mixed, precipitation occurs until a solution equilibrium (a saturated solution) is established. It is possible to simplify these calculations by using the 'bounce back' technique.

It is assumed that during the first change (Δ_1) the system is a reaction that goes to completion, rather than an equilibrium reaction. The two solutions react until one of the ions is completely used up. Then, at the second change Δ_2, the system 'bounces back' to the condition of solubility equilibrium.

Example

2. Lead iodate, $Pb(IO_3)_2$, is a sparingly soluble salt with a K_{sp} of 2.6×10^{-13} at 25°C. To 35.0 mL of 0.150 M $Pb(NO_3)_2$ solution is added 15.0 mL of 0.800 M KIO_3. A precipitate of $Pb(IO_3)_2$ results.

What are the concentrations of Pb^{2+} and IO_3^- in the final solution? Assume the volumes of the solutions are additive.

$[Pb^{2+}] = \dfrac{35.0 \text{ mL} \times 0.150 \text{ mol L}^{-1}}{50.0 \text{ mL}} = 0.105 \text{ M}$

$[IO_3^-] = \dfrac{15 \text{ mL} \times 0.800 \text{ mol L}^{-1}}{50.0 \text{ mL}} = 0.240 \text{ M}$

Let x = amount of Pb^{2+} formed during 'bounce back'.

	$Pb(IO_3)_2(s) \Leftrightarrow$	$Pb^{2+}(aq)$	$+ 2IO_3^-(aq)$
Start	0	0.105 M	0.240 M
Δ_1	+0.105	-0.105	-0.210
Precipitate	Some	0	0.030
Δ_2	-x	+x	+2x
Finish	Some- x	x	0.030
Note: **2x** is small compared to 0.030, and is ignored.			

$K_{sp} = [Pb^{2+}][IO_3^-]^2 = [x][0.030]^2 = 2.6 \times 10^{-13}$

$[Pb^{2+}] = x = \mathbf{2.9 \times 10^{-10} \text{ M}}$ $\qquad\qquad [IO_3^-] = \mathbf{0.030 \text{ M}}$

9. Strong Acids and Strong Bases

Strong acids and bases are completely ionized in water solution. $HClO_4$, HI, HBr, HCl, and HNO_3 are strong acids, and react with water to give solutions with a hydronium ion, $H_3O^+(aq)$, and a concentration equal to that of the initial acid. The hydronium ion is abbreviated as $H^+(aq)$.

1. Example

A solution is prepared by dissolving 0.10 mole of HCl(g) in 1.0 L of water. What is the concentration of ions in the resulting solution?

	HCl(g)	+ H$_2$O(l)	⇒ H$^+$(aq)	+ Cl$^-$(aq)
Start	0.10 M	Some	0.0 M	0.0 M
Δ	0.10	-0.10	+0.10	+0.10
Finish	0.0	Some	0.10	0.10

The Group 1A hydroxides (NaOH, KOH, etc.) are strong bases, and dissociate completely, giving a hydroxide ion molarity, [OH$^-$], equal to that of the starting ionic species.

2. Example

A solution is prepared by dissolving 0.10 mole of NaOH(s) in 1.0 L of water. What is the concentration of ions in the resulting solution?

	NaOH(s)	⇒ Na$^+$(aq)	+ OH$^-$(aq)
Start	0.10 M	0.0 M	0.0 M
Δ	-0.10 M	+0.10 M	+0.10 M
Finish	0.0 M	0.10 M	0.10 M

Water ionizes only slightly and the water equilibrium constant, K_w, is 1.0×10^{-14}.
The equilibrium law equation for water is:
$$K_w = [H^+][OH^-] = 1.0 \times 10^{-14}$$

Equilibrium

3. **Example**

(a) What is the concentration of ions in pure water?

	$H_2O(l)$	$\Leftrightarrow H^+(aq)$	$+ OH^-(aq)$
Start	Some	0.0 M	0.0 M
Δ	$-x$	$+x$	$+x$
Finish	Some	x	x

$[H^+][OH^-] = x^2 = 1.0 \times 10^{-14}$

At equilibrium:
$x = [H^+] = [OH^-] = 1.0 \times 10^{-7}$

(b) What is the pH of water?
The concentration of $H^+(aq)$ in a solution is often expressed using the pH scale. The pH is defined as: $pH = -\log[H^+]$
$pH = -\log[H^+] = 7$

4. **Examples**

What is the solution pH in examples 1, 2, and 3?

Example 1: $[H^+]$ = 0.10 M = 1.0×10^{-1} M
pH = **1.00**

Example 2: $[OH^-]$ = 0.10 M = 1.0×10^{-1} M
K_w = $[H^+][OH^-]$ = 1.0×10^{-14}

$[H^+] = \dfrac{1.0 \times 10^{-14}}{1.0 \times 10^{-1}} = 1.0 \times 10^{-13}$

pH = **13.00**

Example 3: $[H^+]$ = 1.0×10^{-7} M
pH = **7.00**

Example

5. What is the pH corresponding to a hydrogen ion concentration, $H^+(aq)$, of 5.00×10^{-4}? The log of this number is conveniently found with a calculator.

$pH = -\log[H^+] = -\log(5.00 \times 10^{-4}) = \mathbf{3.30}$

6. Find the pH of a solution that results from the mixing of 26.0 mL of 0.200 M KOH with 50.00 mL of 0.100 M HBr.

HBr and KOH completely ionize—they are a strong acid and a stronge base respectively.

$[KOH] = [K^+] = [OH^-] = \dfrac{26.0 \text{ mL} \times 0.200 \text{ mol L}^{-1}}{76.0 \text{ mL}} = 0.0684 \text{ M}$

$[HBr] = [H^+] = [Br^-] = \dfrac{50.0 \text{ mL} \times 0.200 \text{ mol L}^{-1}}{76.0 \text{ mL}} = 0.0658 \text{ M}$

The 'bounce back' technique is used to get the final concentrations using the water equilibrium reaction.

	$H_2O(l)$	$\Leftrightarrow H^+(aq)$	$+ OH^-(aq)$
Start	Some	0.0658 M	0.0684 M
Δ_1	+0.0658	-0.0658 M	-0.0658 M
Intermediate	Some	0	0.0026 M
Δ_2	-x	+x	+x
Finish	Some	x	0.0026
Note: x is small compared to 0.0026, and is ignored.			

$[H^+][OH^-] = [x][0.0026] = 1.0 \times 10^{-14}$
$x = [H^+] = 3.84 \times 10^{-12} \text{ M}$
pH = 11.41

10. Weak Acids, Weak Bases, and pH

Most acids and bases do not completely ionize in water—they are weak. The acid or base reacts with water and is then come to equilibrium with the resulting ions. The most common definition of acids and bases used in AP Chemistry is the Brönsted-Lowry theory. An acid is a proton-donor (H^+ donor) and a base is a proton-acceptor (H^+ acceptor). Water is 'amphiprotic' and can act as either an acid or a base—its role is determined by the acid or base added to it.

The extent of acid-base equilibrium reactions is measured by ionization constants (also called dissociation constants): K_a for acids and K_b for bases. The acid ionization constant, K_a, is determined by measuring the pH of acid solutions and is available as tabulated data.

K_b and K_a are related by the equations:

$$K_w = K_a \times K_b = 1.0 \times 10^{-14}$$

$$K_b = \dfrac{1.0 \times 10^{-14}}{K_a}$$

Equilibrium

There are common elements to the solution of acid-base equilibrium problems.

1. *Always* write the reaction equation.

2. Set up a start...Δ...finish (SΔF) table for the reaction.
   a. If given the pH, H^+ or OH^-:
      Insert values into a (SΔF) table to determine K_a or K_b.
   b. If given K_a or K_b:
      Use the (SΔF) table to solve for 'x' to determine pH, H^+ or OH^-.

Examples

1. A 0.0500 M solution of acetic acid has a pH of 3.03.

(a) What is the value of K_a?
$$CH_3COOH(aq) \Leftrightarrow H^+(aq) + CH_3COO^-(aq)$$

pH = -log $[H^+]$ = 3.03
log$[H^+]$ = –3.03
Use a calculator to find the inverse log of –3.03.
$[H^+] = 9.5 \times 10^{-4}$ M

The solution was prepared from acetic acid alone and therefore the number of acetate ions and hydrogen ions are the same.
$[CH_3COO^-] = [H^+] = 9.5 \times 10^{-4}$ M

Equation	$CH_3COOH(aq)$	$\Leftrightarrow H^+(aq)$	$+ CH_3COO^-(aq)$
Start	0.0500 M	0.0 M	0.0 M
Δ	-9.5×10^{-4}	$+9.5 \times 10^{-4}$	$+9.5 \times 10^{-4}$
Finish	0.049	9.5×10^{-4}	9.5×10^{-4}

Examine the (SΔF) table. There is little difference between the starting and finishing concentrations, since x is small when compared to 0.0500 M. The usual practice is to use the starting concentration of the acid, in this case CH_3COOH, as the equilibrium concentration.

Equation	$CH_3COOH(aq)$	$\Leftrightarrow H^+(aq)$	$+CH_3COO^-(aq)$
Finish	0.0500 M	9.5×10^{-4} M	9.5×10^{-4} m

$$K_a = \frac{[H^+][CH_3COO^-]}{[CH_3COOH]} = \frac{[9.5 \times 10^{-4}][9.5 \times 10^{-4}]}{[0.0500]}$$
$K_a = 1.8 \times 10^{-5}$

(b) What is the K_b for the acetate ion, CH_3COO^-?
Acetate ion is the conjugate base of acetic acid.
The K_a for acetic acid is 1.8×10^{-5}.
$$K_b = \frac{K_w}{K_a} = \frac{1.0 \times 10^{-14}}{1.8 \times 10^{-5}} = 5.6 \times 10^{-10}$$

2. Acetate ion, CH_3COO^-, is the conjugate base of acetic acid, and the K_a for acetic acid is 1.8×10^{-5}.

What is the pH of a 0.100 M solution of acetate ion?

Equation	CH_3COO^-	+ H_2O	$\Leftrightarrow$ CH_3COOH	+ OH^-
Finish	0.100 M	Some	x	x

$$K_b = \frac{[CH_3COOH][OH^-]}{[CH_3COO^-]} = \frac{x^2}{0.100} = 5.6 \times 10^{-10}$$

$x = [CH_3COOH] = [OH^-] = 1.4 \times 10^{-6}$ M

$[H^+] = \dfrac{1.0 \times 10^{-14}}{1.4 \times 10^{-6}} = 1.3 \times 10^{-9}$ M

pH = $-\log[H^+]$ = **8.9**

Each proton of a polyprotic acid is ionized with more difficulty than the previous one. LeChatelier's Principle predicts the H^+ ions from a first ionization will repress the second ionization. The second ionization *can be ignored* in pH calculations. The equilibrium is always written for the loss of *only one proton*.

Example

3. (a) Calculate the pH of a 0.100 M H_2S solution. The equilibria are:
   $H_2S(aq) \Leftrightarrow H^+(aq) + HS^-$ $\qquad K_{a_1} = 1.1 \times 10^{-7}$
   $HS^-(aq) \Leftrightarrow H^+(aq) + S^{2-}$ $\qquad K_{a_2} = 1.2 \times 10^{-13}$

Equation	$H_2S(aq)$	$\Leftrightarrow$ $H^+(aq)$	+ $HS^-(aq)$
Finish	0.100 M	x	x

$$K_{a_1} = \frac{[H^+][HS^-]}{[H_2S]} = \frac{[x][x]}{[0.100]} = 1.1 \times 10^{-7}$$

$x = [H^+] = [HS^-] = 1.0 \times 10^{-4}$ M
pH = $-\log[H^+]$ = **4.00**

(b) What is the $[S^{2-}]$?

Equation	$HS^-(aq)$	$\Leftrightarrow$ $H^+(aq)$	+ $S^{2-}(aq)$
Finish	1.0×10^{-4} M	1.0×10^{-4} M	x

$$K_{a_2} = \frac{[H^+][S^{2-}]}{[HS^-]} = \frac{[1.0 \times 10^{-4}][x]}{[1.0 \times 10^{-4}]} = 1.2 \times 10^{-13}$$

$x = [S^{2-}] = \mathbf{1.2 \times 10^{-13}}$ M

11. Buffers

A buffer is an aqueous solution prepared to contain roughly equal amounts of a weak acid and its conjugate (corresponding) base. The buffer resists a change of pH when H^+ ions from a strong acid or OH^- ions from a strong base are added. The equilibrium buffer system works by removing the added H^+ or OH^- ions by a reaction with an amount of the corresponding base or acid in the system.

The basic equations for buffered solutions for the general reaction,
$$HA \Leftrightarrow H^+ + A^-,$$
are derived from the equilibrium law expression:

(a) $\quad [H^+] = K_a \times \dfrac{[HA]}{[A^-]}$

(b) $\quad pH = pK_a + \log \dfrac{[A^-]}{[HA]}$ $\qquad$ (Henderson-Hasselbach equation)

The $[H^+]$ of a buffer is close to its K_a if the amounts of acid and conjugate base are close to equal. It follows that the pH of a buffer is close to its pK_a ($pK_a = -\log [K_a]$).

Examples

1. A buffer solution contains 1.0 M acetic acid ($K_a = 1.8 \times 10^{-5}$) and 1.0 M sodium acetate. What is the pH of the buffer?
$$CH_3COOH(aq) \Leftrightarrow H^+(aq) + CH_3COO^-(aq)$$
$[CH_3COOH] = 1.0$ M and $[CH_3COO^-] = 1.0$ M

$$pH = pK_a + \log \dfrac{[CH_3COO^-]}{[CH_3COOH]} = 4.74 + \log \dfrac{[1.0]}{[1.0]} \quad = \quad 4.74$$

2. What is the pH of the solution when 0.10 mole of HCl is added to 1.00 L of the buffer prepared in the previous problem?

Use the 'bounce back' technique.

Equation	$CH_3COOH(aq)$	$\Leftrightarrow H^+(aq)$	$+ CH_3COO^-(aq)$
Start	1.00 M	0.10 M	1.00 M
Δ_1	+0.10 M	-0.10 M	-0.10 M
Finish$_1$	1.10 M	0.00 M	0.90 M
Δ_2	$-x$	$+x$	$+x$
Finish$_2$	1.10	x	0.90

Note: x is small compared to 1.10 or 0.90, and is ignored.

$$pH = pK_a + \log \dfrac{[CH_3COO^-]}{[CH_3COOH]} = 4.74 + \log \dfrac{[0.90]}{[1.10]} \quad = \quad 4.65$$

A common way to prepare a buffer is to add to a weak acid half as many moles of a strong base (like NaOH). The resulting buffer will contain equal moles of acid and conjugate base. The acid equilibrium equation can be used, but the base equilibrium equation shows the calculation procedure more clearly.

3. A buffered solution is prepared by adding 20.0 mL of a 0.200 M NaOH solution to 50.0 mL of a 0.100 M CH_3COOH solution. What is the pH of the final solution?

$$[OH^-] = \frac{20.0 \text{ mL} \times 0.200 \text{ mol L}^{-1}}{70.0 \text{ mL}} = 0.0571 \text{ M}$$

$$[CH_3COOH] = \frac{50.0 \text{ mL} \times 0.100 \text{ mol L}^{-1}}{70.0 \text{ mL}} = 0.0714 \text{ M}$$

Equation	CH_3COO^-	$+ H_2O$	$\Leftrightarrow CH_3COOH$	$+ OH^-$
Start	0	Some	0.0714 M	0.0571 m
Δ_1	+0.0571	+0.0571	-0.0571	-0.0571
Finish$_1$	0.0571	Some	0.0143	0.00
Δ_2	-x	-x	+x	+x
Finish$_2$	0.0571	Some	0.0143	x
Note: x is small compared to 0.0571 and 0.0143, and is ignored.				

$$K_b = \frac{[CH_3COOH][OH^-]}{[CH_3COO^-]} = \frac{(0.0143)x}{0.0571} = 5.6 \times 10^{-10}$$

$$x = [OH^-] = 2.2 \times 10^{-9} \text{ M}$$

$$[H^+] = \frac{1.0 \times 10^{-14}}{2.2 \times 10^{-9}} = 4.5 \times 10^{-6} \text{ M}$$

$$pH = -\log[H^+] = 5.35$$

12. Titration and Neutralization

The concentration of an acid or a base can be determined by volumetric analysis. During the titration procedure a solution of known concentration (the standard solution) is added to a known volume of a solution with an unknown concentration. At the equivalence point, the number of moles of hydrogen ion (n_{H^+}) is equal to the number of moles of hydroxide ion (n_{OH^-}) and neutral water is formed, $H^+ + OH^- \Rightarrow H_2O$. The moles of acid, n_{acid}, reacting with the moles of base, n_{base}, at the point of neutralization is determined by the stoichiometry of the reaction equation.

$$H_2SO_4(aq) + 2NaOH(aq) \Rightarrow 2NaSO_4(aq) + 2H_2O(l) \quad n_{acid} = \tfrac{1}{2} n_{base}$$

$$H_3PO_4(aq) + 3NaOH(aq) \Rightarrow 3NaPO_4(aq) + 3H_2O(l) \quad n_{acid} = \tfrac{1}{3} n_{base}$$

Generally *monoprotic* acids and bases such as HCl and NaOH are used so that one mole of acid will react with one mole of base. Since the moles of solute is equal to the molarity times the volume in liters, an equation with the same form used for dilution problems is used for neutralization calculations as well.

$$n_{acid} = \tfrac{1}{1} n_{base}$$
$$m_{base} V_{base} = m_{acid} V_{acid}$$

Equilibrium

Any combination of weak and strong acids or bases can be titrated. The reactions are complete—slightly ionized H_2O is the product of the reaction between H^+ and OH^- even though one reactant may be weak and the other strong. This table summarizes the type of solution that results from the four possible combinations of acids and bases.

Acid	Base	Solution at neutralization point	Example
Strong	Strong	Neutral	$HCl + NaOH$
Weak	Strong	Basic	$CH_3COOH + NaOH$
Strong	Weak	Acidic	$HCl + NH_3$
Weak	Weak	Beyond the scope of the AP Chemistry syllabus	

Examples

1. A titration is performed to standardize a base. It requires 35.4 mL of NaOH to neutralize 10.0 mL of 0.104 M HCl. What is the molarity of the base?

    $M_{base} V_{base} = M_{acid} V_{acid}$
    $M_{base} \times (35.4 \text{ mL}) = (0.104 \text{ mol L}^{-1}) \times (10.0 \text{ mL})$
    $M_{base} = \mathbf{0.0293 \text{ M}}$

2. The same titration is performed, except acetic acid, CH_3COOH, is substituted for the HCl. What is the pH of the resulting solution at the equivalence point?

    $n_{CH_3COOH} = n_{NaOH} = (0.0293 \text{ mol L}^{-1}) \times (0.0354 \text{ L})$
    $n_{CH_3COOH} = 0.00104 \text{ moles}$

    $[CH_3COOH] = [OH^-] = \dfrac{0.00104 \text{ mol}}{0.0454 \text{ L}} = 0.0229 \text{ M}$

Equation	CH_3COO^-	$+ H_2O$	$\Leftrightarrow CH_3COOH$	$+ OH^-$
Start	0	Some	0.0229 M	0.0229 M
Δ_1	+0.0229	+0.0229	-0.0229	-0.0229
Finish$_1$	0.0229	Some	0.00	0.00
Δ_2	$-x$	$-x$	$+x$	$+x$
Finish$_2$	0.0229	Some	x	x
Note: x is small compared to 0.0229, and is ignored.				

$K_b = \dfrac{[CH_3COOH][OH^-]}{[CH_3COO^-]} = \dfrac{x^2}{0.0229} = 5.6 \times 10^{-10}$

$x = [OH^-] = 1.6 \times 10^{-4} \text{ M}$

$[H^+] = \dfrac{1.0 \times 10^{-14}}{1.6 \times 10^{-4}} = 6.3 \times 10^{-11} \text{ M}$

$\text{pH} = -\log[H^+] = \mathbf{10.20}$

13. The Titration of a Weak Acid with a Strong Base

There are several common and useful acid-base titrations calculation principles that are reviewed by examining in detail the titration of a weak acid with a strong base. The reaction equation used to examine this titration is the *base* equation.

$$A^-(aq) + H_2O \Leftrightarrow HA(aq) + OH^-(aq)$$

1. The starting solution at the beginning of the titration is a weak acid, HA.
2. At the half-way point of the titration—when *half* the base needed to reach the equivalence point has been added—a buffer is formed. The pH of the solution is *equal* to pK_a. This data is used to calculate K_a.
3. At the *equivalence point*:
   a. The originally present moles of the weak acid, HA, can still be found. The number of moles of the added hydroxide ion, OH^-, is equal to that of the weak acid.
   b. The molarity of the weak acid, HA, can be computed from the molarity and volume of the base used and the beginning volume of the acid.
   $$m_{base}V_{base} = m_{acid}V_{acid}$$
   c. Dilution calculations of the [HA] and [OH^-] and a (SΔF) table are used to compute the molarity of the conjugate base, [A^-], formed during the titration—the [A^-] is the same as the diluted [HA]. 'Bounce back' and K_b allow the pH of the basic solution at the equivalence point to be computed.

The graph shows the change in pH during the titration of a weak acid with a strong base.

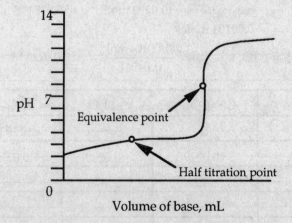

Example
A 0.700 gram sample of an unknown weak monoprotic organic acid, HA, is dissolved in sufficient water to make 50.0 mL of solution and titrated with 0.155 M NaOH solution. A pH of 5.85 is recorded after the addition of 12.5 mL of base. The equivalence point is reached after the addition of 25.0 mL of the 0.155 M NaOH.

(a) Calculate the moles of acid in the original sample.
(b) Calculate the molar mass of the acid, HA.
(c) Calculate the value of the ionization constant, K_a, of the acid HA.
(d) What is the concentration of the original 50.0 mL solution of HA?
(e) What is the pH of the solution at the equivalence point?

Solution

(a) Calculate the moles of acid in the original sample.
At the *equivalence* point:
$$n_{HA} = n_{OH^-} = M_{OH^-} V_{OH^-} = 0.155 \times 0.0250 = 3.88 \times 10^{-3} \text{ mol}$$

(b) Calculate the molar mass of the acid, HA.
$$MM = \frac{0.700 \text{ g}}{3.88 \times 10^{-3} \text{ mol}} = 181 \text{ g mol}^{-1}$$

(c) Calculate the value of the ionization constant, K_a, of the acid HA.
$$pH = pK_a + \log\frac{[A^-]}{[HA]}$$
$$pK_a = pH - \log\frac{[A^-]}{[HA]} = 5.85 - \log\frac{[0.0310]}{[0.0310]} = 5.85$$
$$K_a = \text{Inverse log}[-5.85] = 1.4 \times 10^{-6}$$

(d) What is the concentration of the original 50.0 mL solution of HA?
$$M_{HA} = \frac{M_{OH^-} V_{OH^-}}{V_{HA}} = \frac{0.155 \times 25.0}{50.0} = 0.0775 \text{ M}$$

(e) What is the pH of the solution at the equivalence point?
$$[HA] = \frac{50.0 \times 0.0775}{75.0} = 0.0517 \text{ M}$$
$$[OH^-] = \frac{25.0 \times 0.155}{75.0} = 0.0517 \text{ M}$$

Equation	A^-	$+ H_2O$	$\Leftrightarrow HA$	$+ OH^-$
Start	0	Some	0.0517 M	0.0517 M
Δ_1	+0.0517	+0.0517	-0.0517	-0.0517
Finish$_1$	0.0517	Some	0.00	0.00
Δ_2	$-x$	$-x$	$+x$	$+x$
Finish$_2$	0.0517	Some	x	x
Note: x is small compared to 0.0517, and is ignored.				

$$K_b = \frac{1.0 \times 10^{-14}}{K_a} = \frac{1.0 \times 10^{-14}}{1.4 \times 10^{-6}} = 7.1 \times 10^{-9}$$

$$\frac{x^2}{0.0517} = 7.1 \times 10^{-9}$$

$$x = [OH^-] = 1.9 \times 10^{-5}$$

$$[H^+] = \frac{1.0 \times 10^{-14}}{[OH^-]} = \frac{1.0 \times 10^{-14}}{1.9 \times 10^{-5}} = 5.3 \times 10^{-10}$$

$$pH = -\log[5.3 \times 10^{-10}] = 9.28$$

Chapter 7
Thermodynamics

1. Calorimetry, Internal Energy, and Enthalpy

The 'First Law of Thermodynamics' states that "energy can be converted from one form into another but cannot be created or destroyed." The energy is either in the form of heat, q, or work, w. Heat energy can be converted into work, and work (with systems involving gases) can be converted into heat. The system does not contain heat or work. Heat and work are ways in which energy is *transferred* into or out of the system.

The transfer of work, w, can involve the action of a force that causes a change in the volume of a constant pressure system. This is, most importantly, the compression or expansion of a system that contains gases.

Energy transfer as heat, q, occurs when there is a difference in temperature between the system and the surroundings. The heat transfer always occurs from a region of high temperature to a region of low temperature.

This table gives the most widely used sign convention for heat and work. The system shown is consistent with other thermodynamic computations.

	Work, w	Heat, q
Done *on* a system	+w	+q (endothermic)
Done *by* a system	–w	–q (exothermic)

The thermodynamic 'state' functions, internal energy change, ΔE, enthalpy change, ΔH, entropy change, ΔS and free energy change, ΔG, are independent of how a chemical system comes to a 'state'. These state functions, and their numerical values, are found as data in chemical references. 'Thermo' is concerned with the changes in state functions during chemical or physical changes. Thermodynamic data may be used to make predictions about a specified chemical reaction. The possible predictions include:

1. the heat of reaction (ΔH)
2. whether the reaction is spontaneous ($\Delta G < 0$)
3. the value of the equilibrium constant, K

INTERNAL ENERGY change, ΔE, is a measure of the change of total energy of the system by the application of heat and/or work. Using the convention that heat or work done on the system is absorbed, and heat or work done by the system is evolved:

$$\Delta E = q + w$$

HEAT, q, is measured in joules
(The specific heat of water = 4.184 J g^{-1} °C^{-1}).
q (in joules) = $4.184 \times m_{H_2O} \times \Delta t_{H_2O}$ (for H$_2$O and dilute aqueous solutions)

Thermodynamics

Pressure-volume **WORK** should be converted from atmospheres and liters into joules so it can be added or subtracted from heat.

$$w = P\Delta V = \Delta n_{gas}RT$$
Use $R = 8.314$ J mol^{-1} K^{-1}

To convert work in liter atm to joules multiply by 101:

$$\text{Work} = \frac{R(\text{in J})}{R(\text{in L atm})} = \frac{8.314 \text{ J mol}^{-1} \text{ K}^{-1}}{0.0821 \text{ Liter atm mol}^{-1} \text{ K}^{-1}}$$

Work = 101 J Liter^{-1} atm^{-1}

Examples

1. A gas absorbs 300. J of heat energy and is compressed from 20.0 L to 10.0 L by an opposing pressure of 2.00 atm.
   What is the ΔE for this process, in joules?

   $\Delta E = q + w = q + P\Delta V$
   $\Delta E = +300$ J $+ (2.00$ atm $\times 10.0$ L $\times 101$ J Liter^{-1} atm^{-1})
   Answer: $\Delta E = $ **+2330 J**

ENTHALPY CHANGE, ΔH, is a measure of the change of heat content of a system. It is defined in terms of ΔE.

$\Delta H = \Delta E + w = \Delta E + (\Delta n)RT$
Δn is the change in moles of gases during the reaction.
$\Delta n =$ (number of *gaseous* moles of product – number of *gaseous* moles of reactant).

2. The ΔE for the sublimation of 1 mol of iodine is +59.9 kJ at 25°C and 1.00 atm. What is ΔH for the reaction?
   $$I_2(s) \Leftrightarrow I_2(g)$$

   $\Delta n = (1 - 0) = +1$
   $\Delta H = \Delta E + (\Delta n)RT$
   $\Delta H = +59.9$ kJ $+ ((+1) \times 8.314$ J mol^{-1} K$^{-1} \times 298$ K)
   Answer: $\Delta H = $ **+62.4 kJ**

The internal energy change, ΔE can be determined from the amount of heat evolved in a constant volume calorimeter (a 'bomb' calorimeter). Most frequently, as in an AP Chemistry experiment, heat is determined in a calorimeter at constant atmospheric pressure. The device, often a covered styrofoam cup, will give a direct measure of enthalpy, ΔH.

$\Delta E = q_v$ (in a constant volume calorimeter—a 'bomb' calorimeter)
$\Delta H = q_p$ (in a constant pressure calorimeter—a styrofoam cup)

3. A calibrated bomb calorimeter takes 1500. joules to raise the temperature of the device and its contents 1.000°C. (The heat capacity of the calorimeter is 1500. J °C^{-1}).
A temperature rise of 3.908°C is measured when the calorimeter is used in the combustion of a 1.000 g sample of powdered terbium, Tb, in excess oxygen.

$$4Tb(s) + 3O_2(g) \Rightarrow 2Tb_2O_3(s)$$

What is the internal energy change, ΔE, for the reaction in kJ mol^{-1} of Tb_2O_3.

Since there is no change in volume:
$q_v = 1500.\ J\ °C^{-1} \times 3.908°C = -5862\ J$ (negative sign—*exothermic* reaction)

$n_{Tb} = \dfrac{1.000\ g}{158.9\ g\ mol^{-1}} = 0.006293\ mol$

$n_{Tb_2O_3} = ½\ n_{Tb} = 0.003147\ mol$

$\Delta E = \dfrac{5562\ J}{0.003147\ mol} = -1{,}863{,}000\ J\ mol^{-1}$

Answer: $\Delta E = -1{,}863\ kJ\ mol^{-1}$

4. A 0.25 mol sample of crystalline ammonium chloride, NH_4Cl, is dissolved in 500. mL of pure water in an insulated container at atmospheric pressure and 21.4°C. The solute dissolved with no change in the volume, but the temperature of the solution decreased to 17.8°C. The density and specific heat capacity of the solution is the same as pure water, 1.00 g cm^{-3} and 4.184 J g^{-1} °C^{-1}, respectively.

$$NH_4Cl(s) \Rightarrow NH_4^+(aq) + Cl^-(aq)$$

What is the value of $\Delta H_{solution}$ for NH_4Cl (in kJ mol^{-1})?

$q = 4.184 \times m_{H_2O} \times \Delta t_{H_2O}$ (in joules)
Since there is no change in pressure q is q_p:
$q_p = 4.184 \times 500.\ g \times 3.6°C = +7500\ J$ (positive sign—*endothermic* reaction)
$q_p = \Delta H$ (in J mol^{-1})

$\Delta H = \dfrac{+7500\ J}{0.25\ mol} = +30{,}000\ J\ mol^{-1}$

Answer: $\Delta H = +30.\ kJ\ mol^{-1}$

2. Hess' Law and Heat of Reaction

Hess' Law states that the ΔH for any reaction is a constant, which implies that the heats of reaction are additive. The overall heat of reaction is the algebraic sum of the heats of reaction for the reactions that add up to the that of the overall reaction.

Enthalpy changes slightly with temperature and pressure, but changes dramatically with changes in state. In order for a series of reactions to be added, the states under which the heats of reaction are measured must be known. The heat of reaction where all substances are in their 'standard states' at 25°C and 1.0 atmosphere is written with a superscript, $\Delta H°$. For example, the standard state for water is the liquid state, not the gaseous or the solid state. Oxygen, O_2, is a gas at 25°C and 1.0 atm, while iron, Fe, is a solid.

When working with Hess' Law:

a. The overall reaction is used as a guide to manipulate the steps so that they will add up to the overall reaction. A substance on the product side of an equation should be cancelled by an equal number of moles of that substance on the reactant side of the next equation.

b. Whatever is done to the reaction is reflected in $\Delta H°$. If the reaction is doubled, $\Delta H°$ is doubled. If the reaction is reversed, the sign of $\Delta H°$ should be reversed.

c. When the steps add up to the overall reaction, the sum of each $\Delta H°$ should equal the overall $\Delta H°$.

Example

1. Find ΔH, in kJ, for the reaction from the standard heat of reaction data given.

    $2XO_2(s) + CO(g) \Rightarrow X_2O_3(s) + CO_2(g)$

Step	Reaction	$\Delta H°$
1	$XO_2(s) + CO(g) \Rightarrow XO(s) + CO_2(g)$	–26.8 kJ
2	$X_3O_4(s) + CO(g) \Rightarrow 3XO(s) + CO_2(g)$	+7.3 kJ
3	$3X_2O_3(s) + CO(g) \Rightarrow 2X_3O_4(s) + CO_2(g)$	–10.6 kJ

Step	Action taken on step	Reaction	$\Delta H°$
1	doubled	$2XO_2 + 2CO \Rightarrow \cancel{2XO} + 2CO_2$	–53.6 kJ
2	reversed and ⅔ taken	$\cancel{2XO} + ⅔ CO_2 \Rightarrow ⅔\cancel{X_3O_4} + ⅔ CO$	–4.9 kJ
3	reversed and ⅓ taken	$⅔\cancel{X_3O_4} + ⅓ CO_2 \Rightarrow X_2O_3 + ⅓ CO$	+3.5 kJ
Total		$2XO_2 + 2CO + CO_2 \Rightarrow X_2O_3 + 2CO_2 + CO$	
The desired overall reaction results when the equation is simplified by combining the reactant and product amounts of CO and CO_2.			
Overall		$2XO_2 + CO \Rightarrow X_2O_3 + CO_2$	–55.0 kJ

3. Heats of Formation, Combustion, and Reaction

The standard molar heat of formation, $\Delta H_f°$, is the enthalpy change for the reaction where 1.0 mole of compound in its standard state is made from its elements in their standard states at 25°C.

$\Delta H_f°$ of C_3H_8 (g)	$3C(s) + 4H_2(g) \Rightarrow C_3H_8(g)$
$\Delta H_f°$ of $Fe_3O_4(s)$	$3Fe(s) + 2O_2(g) \Rightarrow Fe_3O_4(s)$

The heat of reaction, $\Delta H°$, can be calculated from the heat of formation of the compounds involved in the reaction. The heat of formation of all elements is zero.

$$\Delta H° = \Sigma \Delta H_f° \text{ (Products)} - \Sigma \Delta H_f° \text{ (Reactants)}$$

The standard molar heat of combustion, $\Delta H_c°$, is the enthalpy change for the reaction where 1.0 mole of compound is burned in oxygen to form carbon dioxide and water, with all substances being in their standard states at 25°C.

$\Delta H_c°$ of $CH_3OH(l)$	$CH_3OH(l) + \frac{3}{2} O_2(g) \Rightarrow CO_2(g) + 2H_2O(l)$
$\Delta H_c°$ of C_3H_8 (g)	$C_3H_8(g) + 5O_2(g) \Rightarrow 3CO_2(g) + 4H_2O(l)$

An additional method of calculating heat of reaction would be from standard heat of combustion data. The sum of the heats of combustion in reactions is additive, as predicted by Hess' Law.

Examples

1. What is the heat of reaction, $\Delta H°$, in kJ mol^{-1} for the reaction?

    $$NH_3(g) + 3F_2(g) \Rightarrow NF_3(g) + 3HF(g)$$

    The standard heats of formation, $\Delta H_f°$, are listed in the table under each substance.

	$NH_3(g)$	$+ 3F_2(g)$	$\Rightarrow NF_3(g)$	$+ 3HF(g)$
$\Delta H_f°$ kJ mol^{-1}	–46.1	0	–125	–271

    $\Delta H° = \Sigma \Delta H_f° \text{ (Products)} - \Sigma \Delta H_f° \text{ (Reactants)}$
    $\Delta H° = ((1 \times -125) + (3 \times -271)) - ((1 \times -46.1) + (3 \times 0))$
    Answer: $\Delta H° = -892$ kJ

2. For the liquids C_2H_5OH and C_2H_4O, the standard heats of combustion are given in the table.
Find $\Delta H°$, in kJ, for the partial oxidation:
$$2C_2H_5OH(l) + O_2(g) \Rightarrow 2C_2H_4O(l) + 2H_2O(l)$$

	Heat of combustion reactions:	$\Delta H_c°$
1	$C_2H_5OH(l) + 3O_2(g) \Rightarrow 2CO_2(g) + 3H_2O(l)$	–1413 kJ
2	$C_2H_4O(l) + {}^5/_2 O_2(g) \Rightarrow 2CO_2(g) + 2H_2O(l)$	–1204 kJ

Step	Action taken on step	Reaction	$\Delta H°$
1	doubled	$2C_2H_5OH(l) + 6O_2(g) \Rightarrow 4CO_2(g) + 6H_2O(l)$	–2826 kJ
2	reversed and doubled	$4CO_2(g) + 4H_2O(l) \Rightarrow 2C_2H_4O(l) + 5O_2(g)$	+2408 kJ
Overall		$2C_2H_5OH(l) + O_2(g) \Rightarrow 2C_2H_4O(l) + 2H_2O(l)$	–418 kJ

Answer: –418 kJ

4. Bond energy

Bond energy is a measure of the strength of covalent bonds. When a bond is broken the potential energy (enthalpy) of the bond is increased. The reaction is endothermic ($\Delta H° > 0$ or $\Delta H°$ is +). When a bond is formed the bond distance is decreased and the reaction is exothermic ($\Delta H° < 0$ or $\Delta H°$ is –).

Bond energies can be used to determine the heat of reaction. The *algebraic* sum of the heat absorbed (+ heat) during the breaking of the bonds of the reactants and the heat released (– heat) during the formation of products gives the heat of reaction. Another type of problem uses the given heat of reaction, $\Delta H°$, data to determine bond energy.

Example
The standard heat of formation of NF_3 is –125 kJ mol^{-1}. The bond energies for N_2 and F_2 are given in the table. Find the average energy for an NF bond, in kJ mol^{-1}.

	$N_2(g)$	$+ 3F_2(g)$	$\Rightarrow 2NF_3(g)$
Bond energy, kJ mol^{-1}	941	155	6'X'

Heat of bond breaking = (941) + (3 × 155) = +1406 kJ
Heat of bond formation = –6'X' kJ
$\Delta H°$ = Bond breaking + Bond formation:
$\Delta H°$ = –125 kJ/mol = (+1406) + (–6'X'))
Answer: NF bond energy = 'X' = **255 kJ mol^{-1}**

5. Entropy

There are two driving forces for chemical reactions. One is enthalpy change, ΔH, which restricts spontaneous reactions to a state of minimum energy. An exothermic reaction, with a ΔH less than zero, is more likely to be spontaneous than an endothermic reaction.

The other driving force is entropy, S. Entropy is a measure of randomness or chaos. The 'second law of thermodynamics' restricts spontaneous reactions to the direction of maximum chaos. Processes with an entropy change, ΔS, greater than zero have increased in chaos, disorder, or randomness and are more likely to occur.

The third law of thermodynamics defines minimum entropy and maximum order. The absolute entropy is zero only for pure crystalline solids at absolute zero (0 K). On the basis of this third law, absolute entropy values can be calculated.

$$S° = \int c_p \, dT = q_p/T$$

Entropy changes can be calculated from absolute entropy using a procedure similar to that used for enthalpy change.

$$\Delta S° = \Sigma S°_{products} - \Sigma S°_{reactants}$$

Examples

1. Which will have a ΔS less than zero (ΔS is negative, –)?

    (A) $2H_2O(g) \Rightarrow 2H_2(g) + O_2(g)$
    (B) $H_2O(g) \Rightarrow H_2O(l)$
    (C) $CaCO_3(s) \Rightarrow CaO(s) + CO_2(g)$
    (D) $2NH_3(g) \Rightarrow N_2(g) + 3H_2(g)$
    (E) none of these

    Answer: (B)
    Liquid water is less chaotic than gaseous water, so the randomness is decreasing in the reaction.

2. Determine the entropy change at 25°C, in $J\,K^{-1}$ for:

	$2SO_2(g)$	$+ O_2(g)$	$\Rightarrow 2SO_3(g)$
$S°$ ($J\,K^{-1}\,mol^{-1}$)	248.1	205.03	256.6

    $\Delta S° = \Sigma S°_{products} - \Sigma S°_{reactants}$
    $\Delta S° = (2 \times 256.6) - ((2 \times 248.1) + (1 \times 205.03))$

    Answer: $\Delta S° = -188.0 \; J\,K^{-1}$

6. Free Energy

The standard free energy change, $\Delta G°$, for a reaction at constant temperature and pressure is defined as:

$$\Delta G° = \Delta H° - \frac{T\Delta S°}{1000}$$

The $T\Delta S°$ term is divided by 1000 because the units of entropy are joules, while the units of free energy and enthalpy are *kilo*joules.

The free energy change, or net driving force, is the solution to an equation which contains the two driving forces for chemical reactions, enthalpy and entropy change. It is possible to determine reaction spontaneity from the addition of the two terms. The result depends on the value and the sign of $\Delta H°$ and $T\Delta S°$.

If $\Delta G°$ is negative (–), the reaction *is* spontaneous.
If $\Delta G° = 0$, the reaction is at equilibrium.
If $\Delta G°$ is positive (+), the reaction *is not* spontaneous.

The relationship between $\Delta G°$, $\Delta H°$ and $T\Delta S°$.

Process	$\Delta H°$	$T\Delta S°$	$\Delta G° = \Delta H° - T\Delta S°$
1	–	+	–
2	+	–	+
3	+	+	?
4	–	–	?

- Process 1 is definitely *spontaneous*. (ΔG is negative, –.)
  (Both ΔH and ΔS are favorable for spontaneous reaction.)

- Process 2 is definitely *nonspontaneous*. (ΔG is positive, +.)
  (Both ΔH and ΔS are unfavorable for spontaneous reaction.)

- Processes 3 and 4 are possibly spontaneous. (ΔG is questionable, ?.)

  Process 3 can occur at *high* temperatures.
  ($T\Delta S$ must be **large**. ΔH is unfavorable. $T\Delta S$ is favorable.)

  Process 4 can occur at *low* temperatures.
  ($T\Delta S$ must be **small**. ΔH is favorable. $T\Delta S$ is unfavorable.)

The standard free energy change of formation of a compound, $\Delta G_f°$, is the change in net driving force for a reaction at 25°C. The reactants are elements in their standard states. One (1.0) mole of the compound is formed in its standard state. The free energy of formation, $\Delta G_f°$, of elements is zero, just as it was for the enthalpy of formation, $\Delta H_f°$, of elements. The free energy change of a reaction, $\Delta G°$, is calculated from the free energy of formation of the compounds involved in a reaction using the equation:

$$\Delta G° = \sum \Delta G_f° \text{ (Products)} - \sum \Delta G_f° \text{ (Reactants)}$$

Examples

1. Estimate the approximate temperature, in °C, at which the reaction given will become spontaneous.

$$CaCO_3(s) \Rightarrow CaO(s) + CO_2(g)$$

The reaction occurs at 25°C, is endothermic, and has a positive entropy change. $\Delta H°$ is +184 kJ, $\Delta S°$ is +166 J K^{-1} and $\Delta G°$ is + 300. kJ. It can be assumed that ΔH and ΔS will not vary in this range.

The reaction goes from unfavorable to favorable when $\Delta G = 0$.

$$\Delta G° = \Delta H° - \frac{T\Delta S°}{1000}$$

$$\frac{T\Delta S°}{1000} = \Delta H°$$

$$T = \frac{(1000) \times (+184 \text{ kJ})}{+166 \text{ J K}^{-1}} = 1108 \text{ K} = 835°C$$

Answer: The temperature must be *greater* than **835°C**.

2. Calculate $\Delta G°$ for the reaction:

	4NH$_3$(g)	+ 5O$_2$(g)	$\Rightarrow$ 4NO(g)	+ 6H$_2$O(g)
$\Delta G_f°$ kJ mol^{-1}	−16.48	0	86.67	−228.59

$\Delta G°$ = $\Sigma \Delta G_f°$ (Products)   − $\Sigma \Delta G_f°$ (Reactants)
$\Delta G°$ = $((4 \times 86.67) + (6 \times -228.59))$ − $((4 \times -16.48) + (0))$
Answer: $\Delta G° = -958.8$ kJ

7. Free Energy and Equilibrium

An important use of free energy change is to determine values of the equilibrium constant, K. ΔG is the free energy change for a reaction containing substances which are *not* in their standard state. The equation relating ΔG, $\Delta G°$ and the equilibrium constant, K is:

$$\Delta G = \Delta G° + RT \ln K$$

At equilibrium, ΔG = zero (0), and the equation becomes:

$$\Delta G° = -RT \ln K$$

Thermodynamics

Examples

1. For the reaction:

   $$2CO(g) + O_2(g) \Leftrightarrow 2CO_2(g)$$

   The $\Delta G°$ for the reaction is -257.2 kJ mol^{-1}.
   What is the equilibrium constant at 25 °C?

   $\Delta G°$ = $(-275.2$ kJ$) \times (1000$ J kJ$^{-1}) = -275,200$ J
   $\Delta G°$ = $-RT \ln K$
   $-275,200$ J = $-(8.314$ J mol^{-1} K$^{-1}) \times (298$ K$) \times \ln K$
   $\ln K$ = 111.1

   Answer: $K_p = 2 \times 10^{48}$
   This is K_p since the gaseous state is standard for all of the chemicals in the reaction.

2. $\Delta G° = -42.51$ kJ for the reaction at 298 K:
   $$Ag^+(aq) + 2NH_3(aq) \Leftrightarrow Ag(NH_3)_2^+(aq)$$
   What is the equilibrium constant, K_c?

   $\Delta G°$ = $(-42.51$ kJ$) \times (1000$ J kJ$^{-1}) = -42,510$ J
   $\Delta G°$ = $-RT \ln K$
   $-42,510$ J = $-(8.314$ J mol^{-1} K$^{-1}) \times (298$ K$) \times \ln K$
   $\ln K$ = 17.16

   Answer: $K_c = 2.8 \times 10^7$
   This is K_c since concentration, the molarity of an aqueous solution, is standard for this reaction.

3. Calculate the equilibrium vapor pressure of water at 25°C.
   $$H_2O(l) \Leftrightarrow H_2O(g)$$
   $\Delta G_f° = -244.6$ kJ mol^{-1} for $H_2O(l)$
   $\Delta G_f° = -235.8$ kJ mol^{-1} for $H_2O(g)$

   K_p = $P_{H_2O(g)}$ = the vapor pressure of H_2O
   $\Delta G°$ = $-RT \ln K_p$
   $\Delta G°$ = $(-235.8) - (-244.6) = +8.8$ kJ $= 8,800$ J
   $8,800$ J = $-(8.314$ J mol^{-1} K$^{-1}) \times (298$ K$) \times \ln K$
   $\ln K$ = -3.55

   Answer: $K_p = P^°_{H_2O} = 0.028$ atm $\times 760$ mmHg atm^{-1} = 21.8 mmHg
   K_p is measured in atmospheres. This is the standard unit of pressure.

8. Free Energy Change and Net Cell Potential

The further an oxidation-reduction (redox) reaction is away from equilibrium, the greater will be the voltage or cell potential. The net driving force (free energy change) of a redox reaction is related to cell potential by the equation:

$\Delta G° = -nF\mathcal{E}°$

where:
F = Faradays
F = 96,500 J mol^{-1} V^{-1}

$\mathcal{E}°$ = Standard cell potential, volts (V)

Examples

1. Consider the reaction:
$$Mg(s) + 2H^+(1\,M) \Leftrightarrow Mg^{2+}(1\,M) + H_2(1.0\,atm)$$
The standard free energy, $\Delta G°$, at 25°C for this reaction is –468.9 kJ mol^{-1}? What is the standard cell potential, $\mathcal{E}°$, for this reaction at 25°C?

$\Delta G°$ = (–468.9 kJ) × (1000 J kJ^{-1}) = –468,900 J
$\Delta G°$ *must* be in joules to match the units of F, 96,500 J V^{-1}.
n = 2 moles of electrons are transferred from Mg to the 2H$^+$.
$\Delta G°$ = $-nF\mathcal{E}°$

–468,900 J = – (2 mol) × (96,500 J V^{-1}) × $\mathcal{E}°$

Answer: $\mathcal{E}°$ = 2.430 V

2. The standard reduction potential for V^{2+} forming V(s) is –1.175 V. The standard reduction potential, $\mathcal{E}°$, for TeO$_2$ forming Te is +0.593 V. Calculate the standard free energy change, $\Delta G°$, in kJ, for the reaction:
$$V(s) + TeO_2(s) + 4H^+(aq) \Leftrightarrow V^{2+}(aq) + Te(s) + 2H_2O(l)$$

Reaction	Equation	$\mathcal{E}°$, V
Oxidation:	$V(s) \Leftrightarrow V^{2+}(aq) + 2e^-$	+1.175
Reduction:	$TeO_2(s) + 4H^+(aq) + 2e^- \Leftrightarrow Te(s) + 2H_2O$	+0.593
Net Cell Potential:	$V(s) + TeO_2(s) + 4H^+(aq) \Leftrightarrow V^{2+}(aq) + Te(s) + 2H_2O(l)$	+1.768

$\Delta G°$ = $-nF\mathcal{E}°$
n = 2 moles of electrons are transferred from V to TeO$_2$.
$\Delta G°$ = –(2.00 mol) × (96,500 J V^{-1}) × (1.768 V)

Answer: $\Delta G°$ = –341,000 J = –341 kJ

Chapter 8
Electrochemistry

1. **Balancing Oxidation-Reduction Reactions**

Reactions in which electrons are transferred are oxidation-reduction or redox reactions. The status of the electron transfer in redox reactions is evaluated by means of oxidation numbers. Every reaction of this type involves at least one element whose oxidation number is increasing and at least one whose oxidation number is decreasing.

Reducing agents:	Oxidizing agents:
are oxidized	are reduced
lose electrons	gain electrons
increase in oxidation number	decrease in oxidation number

The rules for assigning oxidation numbers are:
- The oxidation number of an *element* is 0.
- In *compounds*:

Ion	Oxidation #	Ion	Oxidation #
Hydrogen, H	+1	Hydride	-1
Alkali metal (Group IA)	+1	Binary halides (F, Cl, Br, I)	-1
Alkaline earth (Group IIA)	+2	Oxide, O or Sulfide, S (Group VIA)	-2
Al and Ga (Group IIIA)	+3	Nitride, N (Group VA)	-3

- The *sum of the oxidation numbers*:
  a. in a compound is 0
  b. in an ion is the charge on the ion

The balancing of redox reactions often cannot be accomplished by merely inspecting the number of reactant and product atoms alone. The gain and loss of electrons must also be accounted for in the balancing.

An efficient method of balancing redox reactions involves following these rules.

1. Identify the substance being oxidized and the substance being reduced. Expect that only one element in the reaction is oxidized and one element is reduced. Write an oxidation half-reaction and a reduction half-reaction. Don't worry yet about which is which.

2. Balance each half-reaction with respect to all elements, except hydrogen, H, and oxygen, O.

FOR ACID SOLUTIONS:

3. Add the appropriate number of water molecules, H_2O, to the deficient side to balance the oxygen, O.

4. Add H^+ to the deficient side to balance the hydrogen.

FOR BASIC SOLUTIONS:

3. For each deficient oxygen, O:
   Add two (2) hydroxides, OH^-, to the deficient side.
   Add one (1) water, H_2O, to the other side.

4. For each deficient hydrogen, 'H':
   Add one (1) water, H_2O, to the side that is deficient.
   Add one (1) hydroxide, OH^-, to the other side.

5. Add an appropriate number of electrons to balance the charge on the side of the half-reaction that has an excess positive charge.
   It should be easy to identify which is the oxidation reaction (electrons are on the product side) and which is the reduction reaction (electrons are on the reactant side).

6. Multiply each half-reaction by the appropriate integer so the gain of electrons by the reduction reaction equals the loss of electrons by the oxidation reaction.

7. Add the two half-reactions and combine duplications.

Examples

1. Balance the reaction in *acid* solution:
   $Ag^+ + AsH_3 \Rightarrow Ag + H_3AsO_3$

Step	Action	Equations
1	HALF-REACTIONS:	$Ag^+ \Rightarrow Ag$ $AsH_3 \Rightarrow H_3AsO_3$
2	BALANCE ATOMS, except O and H:	$Ag^+ \Rightarrow Ag$ $AsH_3 \Rightarrow H_3AsO_3$
3	BALANCE OXYGEN with H_2O:	$Ag^+ \Rightarrow Ag$ $AsH_3 + 3H_2O \Rightarrow H_3AsO_3$
4	BALANCE HYDROGEN with H^+:	$Ag^+ \Rightarrow Ag$ $AsH_3 + 3H_2O \Rightarrow H_3AsO_3 + 6H^+$
5	BALANCE THE CHARGE with electrons:	$Ag^+ + e^- \Rightarrow Ag$ $AsH_3 + 3H_2O \Rightarrow H_3AsO_3 + 6H^+ + 6e^-$
6	BALANCE THE ELECTRONS:	$6Ag^+ + 6e^- \Rightarrow 6Ag$ $AsH_3 + 3H_2O \Rightarrow H_3AsO_3 + 6H^+ + 6e^-$
7	COMBINE AND SIMPLIFY: **Answer:** $6Ag^+ + AsH_3 + 3H_2O \Rightarrow 6Ag + H_3AsO_3 + 6H^+$	

2. Balance the reaction in *basic* solution:
$$HS_2O_4^- + CrO_4^{2-} \Rightarrow SO_4^{2-} + Cr(OH)_4^-$$

Step	Action	Equations
1	HALF-REACTIONS:	$HS_2O_4^- \Rightarrow SO_4^{2-}$ $CrO_4^{2-} \Rightarrow Cr(OH)_4^-$
2	BALANCE ATOMS, except O and H:	$HS_2O_4^- \Rightarrow 2SO_4^{2-}$ $CrO_4^{2-} \Rightarrow Cr(OH)_4^-$
3	BALANCE OXYGEN with OH^- and H_2O:	$HS_2O_4^- + 8OH^- \Rightarrow 2SO_4^{2-} + 4H_2O$ $CrO_4^{2-} \Rightarrow Cr(OH)_4^-$
4	BALANCE HYDROGEN with H_2O and OH^-:	$HS_2O_4^- + 9OH^- \Rightarrow 2SO_4^{2-} + 5H_2O$ $CrO_4^{2-} + 4H_2O \Rightarrow Cr(OH)_4^- + 4OH^-$
5	BALANCE THE CHARGE with electrons:	$HS_2O_4^- + 9OH^- \Rightarrow 2SO_4^{2-} + 5H_2O + 6e^-$ $CrO_4^{2-} + 4H_2O + 3e^- \Rightarrow Cr(OH)_4^- + 4OH^-$
6	BALANCE THE ELECTRONS:	$HS_2O_4^- + 9OH^- \Rightarrow 2SO_4^{2-} + 5H_2O + 6e^-$ $2CrO_4^{2-} + 8H_2O + 6e^- \Rightarrow 2Cr(OH)_4^- + 8OH^-$
7	COMBINE AND SIMPLIFY: **Answer:** $HS_2O_4^- + OH^- + 2CrO_4^{2-} + 3H_2O \Rightarrow 2SO_4^{2-} + 2Cr(OH)_4^-$	

A great deal of information is available from the steps used to balance a redox reaction when the result of each is put into words.

"The **OXIDATION** half-reaction for the reaction is:
$$HS_2O_4^- + 9OH^- \Rightarrow 2SO_4^{2-} + 5H_2O + 6e^-$$
and the hydrogen hyposulfite ion, $HS_2O_4^-$, is:

...oxidized

...losing electrons

...the reducing agent

...the oxidation number of each sulfur atom increases from +3 to +6."

"The **REDUCTION** half-reaction for the reaction is:
$$CrO_4^{2-} + 4H_2O + 3e^- \Rightarrow Cr(OH)_4^- + 4OH^-$$
and the chromate ion, CrO_4^{2-}, is:

...reduced

...gaining electrons

...the oxidizing agent

...the oxidation number of each chromium atom decreases from +6 to +3."

2. **The Faraday and Electrolytic Cells**

In an electrolytic cell electrical energy is used to bring about a nonspontaneous electrical change. Electrolysis always results in a redox chemical reaction.

Electrochemical cells, both chemical (voltaic) and electrolytic, consist of an electrolyte and electrodes. An 'electrolyte' is a substance that exists as ions in either the molten (fused) state or in aqueous solution. The 'electrodes' consist of an anode and a cathode which are generally metals or graphite, C, rods. The electrodes conduct electrons to an external circuit which in electrolysis is a source of energy. Electrons *always* flow from the anode to the cathode.

The anode is *always* where oxidation (the loss of electrons) occurs.
  Remember 'AN OX'.
  The oxidation reaction which occurs involves the most spontaneous reaction of the given reducing agents.

The cathode is *always* where reduction (the gain of electrons) occurs.
  Remember the 'RED CAT'.
  The reduction reaction involves the reaction of the most active of the given oxidizing agents.

Inert electrodes that will not take part in the electrolysis reaction are often selected. Common examples are graphite, C, platinum, Pt, gold, Au, stainless steel, or nichrome wire. Anodes which are not inert will oxidize during electrolysis. There are three common electrolytic reactions that occur at the cathode and the anode.

	AT AN INERT **ANODE**: (+ electrode during electrolysis)	AT THE **CATHODE**: (− electrode during electrolysis)
1.	An anion (- ion) is *oxidized* to a nonmetallic molecule.	A cation (+ ion) is *reduced* to a metal atom.
2.	When the anion is: F^-, NO_3^-, or SO_4^{2-} in aqueous solution *it will not be oxidized.* Instead: H_2O is oxidized to $O_2 + H^+$	When the cation is a: Group 1A ion or Group 2A ion in aqueous solution *it will not be reduced.* Instead: H_2O is reduced to $H_2 + OH^-$
3.	With a strong base: OH^- is oxidized to $O_2 + H_2O$	With a strong acid: H^+ is reduced to H_2

Electrochemistry

Example

A 1.0 M solution of KI(aq) is electrolyzed between graphite, C, electrodes: Which is true?

(A) The solution will become more basic.

(B) Oxygen will be evolved at the anode.

(C) Iodine will be formed at the cathode.

(D) Hydrogen will be evolved at the anode.

(E) Potassium will be deposited at the cathode.

Answer: (A)

KI in an aqueous solution is not a strong acid or base.

- At the *anode*:
  I^- is oxidized to I_2.

- At the *cathode*:
  K^+ is a Group 1A ion and is not reduced in aqueous solution.
  Thus: H_2O is reduced to $H_2 + OH^-$.

A **Faraday** is the electrical charge in coulombs on one mole of electrons—96,500 C. An **ampere** is the rate of flow of electrons in coulombs second^{-1} (C s^{-1}).

Examples

1. A current of 10. amperes is passed through molten magnesium chloride for 3.0 hours. How many moles of magnesium metal could be produced by this electrolysis?

$$n_{electrons} = \frac{10\ C}{s} \times \frac{1\ mol\ electrons}{96,500\ C} \times 3.0\ h \times \frac{3600\ s}{1\ h}$$

$n_{electrons}$ = 1.1 moles

At the cathode:
$$Mg^{2+} + 2e^- \Rightarrow Mg$$

Using the 'mole link':
$$n_{Mg} = \tfrac{1}{2} n_{e^-} = \tfrac{1}{2} \times 1.1 = 0.56\ moles$$
Answer: 0.56 moles

2. One-half (0.500) Faraday of electricity is passed through aqueous solutions of the compounds listed in the table. Electrodes are made of the material indicated.
Give the formula of the substance that is produced at each electrode and the amount in moles that is produced.

Answers:
A.

		Anode $2H_2O \Rightarrow O_2 + 4H^+ + 4e^-$ $n_{O_2} = \frac{1}{4} n_{e^-}$		Cathode $Fe^{3+} + 3e^- \Rightarrow Fe$ $n_{Fe} = \frac{1}{3} n_{e^-}$	
Compound	Electrode material	Substance produced	Number of moles	Substance produced	Number of moles
$Fe(NO_3)_3$	stainless steel	O_2	0.125	Fe	0.167

B.

		Anode $Ni \Rightarrow Ni^{2+} + 2e^-$ $n_{Ni} = \frac{1}{2} n_{e^-}$		Cathode $Ni^{2+} + 2e^- \Rightarrow Ni$ $n_{Ni} = \frac{1}{2} n_{e^-}$	
Compound	Electrode material	Substance produced	Number of moles	Substance produced	Number of moles
$NiSO_4$	Ni	Ni^{2+}	0.250	Ni	0.250

C.

		Anode $2H_2O \Rightarrow O_2 + 4H^+ + 4e^-$ $n_{O_2} = \frac{1}{4} n_{e^-}$		Cathode $Au^{3+} + 3e^- \Rightarrow Au$ $n_{Au} = \frac{1}{3} n_{e^-}$	
Compound	Electrode material	Substance produced	Number of moles	Substance produced	Number of moles
$Au(NO_3)_3$	Au	O_2	0.125	Au	0.167

D.

Compound	Electrode material	Anode $Cu \Rightarrow Cu^{2+} + 2e^-$ $n_{Cu} = \frac{1}{2}n_{e^-}$		Cathode $H_2O + 2e^- \Rightarrow H_2 + 2OH^-$ $n_{H_2} = \frac{1}{2}n_{e^-}$	
		Substance produced	Number of moles	Substance produced	Number of moles
Li_2SO_4	Cu	Cu^{2+}	0.250	H_2	0.250

E.

Compound	Electrode material	Anode $2Br^- \Rightarrow Br_2 + 2e^-$ $n_{Br_2} = \frac{1}{2}n_{e^-}$		Cathode $H_2O + 2e^- \Rightarrow H_2 + 2OH^-$ $n_{H_2} = \frac{1}{2}n_{e^-}$	
		Substance produced	Number of moles	Substance produced	Number of moles
$BaBr_2$	nichrome	Br_2	0.250	H_2	0.250

F.

Compound	Electrode material	Anode $2H_2O \Rightarrow O_2 + 4H^+ + 4e^-$ $n_{O_2} = \frac{1}{4}n_{e^-}$		Cathode $H_2O + 2e^- \Rightarrow H_2 + 2OH^-$ $n_{H_2} = \frac{1}{2}n_{e^-}$	
		Substance produced	Number of moles	Substance produced	Number of moles
$Al_2(SO_4)_3$	Pt	O_2	0.125	H_2	0.250

G.

Compound	Electrode material	Anode $2H_2O \Rightarrow O_2 + 4H^+ + 4e^-$ $n_{O_2} = \frac{1}{4}n_{e^-}$		Cathode $Sn^{4+} + 4e^- \Rightarrow Sn$ $n_{Sn} = \frac{1}{4}n_{e^-}$	
		Substance produced	Number of moles	Substance produced	Number of moles
SnF_4	C	O_2	0.125	Sn	0.125

H.

		Anode $Cd \Rightarrow Cd^{2+} + 2e^-$ $n_{Cd} = \frac{1}{2} n_{e^-}$		Cathode $Cd^{2+} + 2e^- \Rightarrow Cd$ $n_{Cd} = \frac{1}{2} n_{e^-}$	
Compound	Electrode material	Substance produced	Number of moles	Substance produced	Number of moles
$Cd(NO_3)_2$	Cd	Cd	0.250	Cd^{2+}	0.250

I.

		Anode $Zn \Rightarrow Zn^{2+} + 2e^-$ $n_{Zn^{2+}} = \frac{1}{2} n_{e^-}$		Cathode $H_2O + 2e^- \Rightarrow H_2 + 2OH^-$ $n_{H_2} = \frac{1}{2} n_{e^-}$	
Compound	Electrode material	Substance produced	Number of moles	Substance produced	Number of moles
K_2SO_4	Zn	Zn^{2+}	0.250	H_2	0.250

3. Voltaic (Chemical) Cells

In a voltaic (chemical) cell electrical energy is produced by a spontaneous oxidation-reduction reaction. This spontaneous reaction will occur whenever two dissimilar metallic electrodes are immersed in an electrolyte and connected to an external circuit.

The electrodes of a voltaic cell have charges which are opposite to those of an electrolytic cell. After all, it is working in an opposite manner. The anode is the negative (-) electrode and the cathode is the positive (+) electrode. The 'RED CAT' and 'AN OX' principles stated earlier are true for voltaic cells, as is the rule for the flow of electrons from the anode to the cathode in the external circuit.

An example of a voltaic cell would be the system shown in the diagram on the next page. The electrodes are zinc and copper, and the electrolytes in each half-cell include Zn^{2+} and Cu^{2+}. In the solution will also be an anion, such as NO_3^- or SO_4^{2-}, which is a required part of the salt used to prepare the electrolyte.

Zinc is the anode, and the oxidation of Zn(s) to Zn^{2+} occurs at the electrode. The reduction of copper ions, Cu^{2+}, to copper occurs at the other electrode, which is the cathode. The electrons move from the anode to the cathode in the external circuit of every voltaic cell.

A 'salt bridge' containing a non-reactive electrolyte completes the circuit. Potassium chloride solution, KCl(aq), is used in this system. The positive ions in the salt bridge move from the anode to the cathode because reduction is depleting the concentration of positive ions at the cathode. The negative ions move from the cathode to the anode to balance the charge since oxidation is increasing the concentration of positive ions.

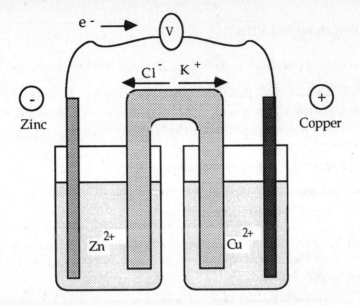

An abbreviation for this cell is:

Zn/Zn^{2+} (Zn)	||	Cu^{2+}/Cu (Cu)
{anode}	{salt bridge}	{cathode}
Zn $\Rightarrow$ Zn^{2+} + 2e$^-$		Cu^{2+} + 2e$^-$ $\Rightarrow$ Cu
{Oxidation}		{Reduction}

Example

Use the abbreviated form of a voltaic cell to show the arrangement for the reaction:
$$Sn^{2+} + Br_2 \Rightarrow Sn^{4+} + 2Br^-$$
The anode is tin, Sn, and the cathode is graphite, C.

Answer:
$$Sn^{2+}/Sn^{4+} (Sn) \; || \; (C) \; Br_2/Br^-$$
Electrons flow through the wire connecting the electrodes from the anode, the tin, Sn, electrode to the cathode, the inert carbon, C, electrode.

Tin was selected as the negative electrode. An inert material (Pt or nichrome) could also have been used.

Anode (Oxidation): $\qquad Sn^{2+} \Rightarrow Sn^{4+} + 2e^-$

Tin(IV) ions, Sn^{4+}(aq), are being formed at the anode. Negative ions will always migrate through the salt bridge from the cathode to the anode to maintain a balance of the charge in the oxidation half-cell.

Graphite was selected as the positive electrode. Bromine or a solution of bromide ion, for example, cannot be electrodes. They are nonconductors of electrons.

Cathode (Reduction): $\qquad Br_2 + 2e^- \Rightarrow 2Br^-$

Bromide ions, Br$^-$ (aq), are being formed at the cathode. Positive ions will always migrate through the salt bridge from the anode to the cathode to maintain a balance of the charge in the reduction half-cell.

4. Standard Reduction Potentials

Tables of standard reduction potentials at 25°C are used to determine cell voltages and to predict whether a redox reaction is spontaneous. The standard state for $\mathcal{E}°$ is the same as it is for the thermodynamic state functions: all substances are in their standard state, gases are at 1.0 atmosphere, and solution concentration is 1.0 M.

The more positive the standard reduction potential the stronger the oxidizing agent and the weaker its conjugate reducing agent. The $\mathcal{E}°$ for an oxidation half-reaction is the standard reduction potential with the opposite sign.

The table of Standard Electrode Potentials shown below conforms to the style used in the references for Section II of the AP examination. It should be noted that this style does *not* conform to that used in some texts.

	Reaction	$\mathcal{E}°$	
Weakest oxidizing agents	$Li^+ + e^- \Rightarrow Li$	- 3.04 volts	**Strongest** reducing agents
	$Zn^{2+} + 2e^- \Rightarrow Zn$	- 0.76 volts	
	$Fe^{2+} + 2e^- \Rightarrow Fe$	- 0.45 volts	
	$Ni^{2+} + 2e^- \Rightarrow Ni$	- 0.26 volts	
	$Sn^{2+} + 2e^- \Rightarrow Sn$	- 0.14 volts	
	$2H^+ + 2e^- \Rightarrow H_2$	0.00 volts	
	$Sn^{4+} + 2e^- \Rightarrow Sn^{2+}$	0.15 volts	
	$Cu^{2+} + 2e^- \Rightarrow Cu$	0.34 volts	
	$I_2 + 2e^- \Rightarrow 2I^-$	0.54 volts	
	$Fe^{3+} + 2e^- \Rightarrow Fe^{2+}$	0.77 volts	
	$Ag^+ + e^- \Rightarrow Ag$	0.80 volts	
Strongest oxidizing agents	$Br_2 + 2e^- \Rightarrow 2Br^-$	1.09 volts	**Weakest** reducing agents
	$Au^{3+} + 3e^- \Rightarrow Au$	1.50 volts	
	$F_2 + 2e^- \Rightarrow 2F^-$	2.87 volts	

The cell voltage is determined by adding the potentials of the oxidation and reduction half-reactions. The half-reaction potential does *not* change when the reaction is multiplied to balance the electrons.

If the cell voltage is positive, the reaction will occur spontaneously. If the voltaic cell has a negative potential, the reverse reaction will be spontaneous. When $\mathcal{E}°$ is 0, the reaction will be at equilibrium in the standard state.

Examples

1. What is the net cell potential and the overall reaction for the voltaic cell?
   The abbreviated form is: Ni/Ni^{2+} || Ag$^+$/Ag

   Anode (Oxidation):

   Ni $\Rightarrow$ Ni^{2+} + 2e$^-$ $\varepsilon° = +0.26$ volts

   For an **oxidation** reaction, the sign of the reduction potential, $\varepsilon°$, is *changed*.

   Cathode (Reduction):

   2Ag$^+$ + 2e$^-$ $\Rightarrow$ 2Ag $\varepsilon° = +0.80$ volts

   The cathode reaction is *doubled* to balance the electrons. The $\varepsilon°$ remains *unchanged*.

   Answer:

   Ni + 2Ag$^+$ $\Rightarrow$ Ni^{2+} + 2Ag $\varepsilon° = +1.06$ volts

2. Which will be oxidized by Ni^{2+}?

   (A) Ag   (D) F$_2$
   (B) Br$^-$ (E) Fe
   (C) Cu

   Answer (E)
   Fe has an oxidation potential (+0.45 volts) high enough to give a positive overall potential.

   Fe $\Rightarrow$ Fe^{2+} + 2e$^-$ $\varepsilon° = +0.45$ volts

   Ni^{2+} + 2e$^-$ $\Rightarrow$ Ni $\varepsilon° = -0.26$ volts

   Fe + Ni^{2+} $\Rightarrow$ Fe^{2+} + Ni $\varepsilon° = +0.19$ volts

3. Which will be reduced by Ag?
   (A) Au^{3+} (D) I$_2$
   (B) Cu   (E) Ni^{2+}
   (C) Fe^{3+}

   Answer: (A)

   Au^{3+} is the only choice with a high enough reduction potential ($\varepsilon° = +1.50$ volts) to oxidize silver, Ag ($\varepsilon° = -0.80$ v), to give a spontaneous overall reaction.

   3Ag + Au^{3+} $\Rightarrow$ 3Ag$^+$ + Au $\varepsilon° = +0.70$ volts

5. The Nernst Equation

An electrochemical cell is a system in the process of achieving chemical equilibrium. The highest voltage occurs when the system is far from equilibrium. LeChatelier's Principle is used to analyze the effect of stresses which are applied or removed. Cell potential is independent of the size of the electrodes, the size of the cells and the volume of electrolyte since the constant concentration of solids and liquids does not affect the kinetics of the equilibrium process. Concentration does affect the voltage. Higher reactant and lower product concentration will increase cell potential.

The Nernst equation is used to determine the voltage of a voltaic cell when the concentrations are not 1.0 M.

(Nernst Equation) at 25°C; $\quad \varepsilon = \varepsilon° - \dfrac{0.0591}{n} \log Q$

Examples

1. The electrolytes of the cell are 0.0500 M $Zn(NO_3)_2$ and 1.000 M $Cu(NO_3)_2$. What will be the initial cell voltage?

    $Zn/Zn^{2+} \;||\; Cu^{2+}/Cu$

    $Zn + Cu^{2+} \Rightarrow Zn^{2+} + Cu \qquad \qquad \varepsilon° = 1.10$ volts

    $\varepsilon = \varepsilon° - \dfrac{0.0592}{n} \log Q =$

    $\varepsilon = \varepsilon° - \dfrac{0.0592}{n} \log \dfrac{[Zn^{2+}]}{[Cu^{2+}]}$

    n = 2 moles of electrons transferred

    $\varepsilon = 1.10 - \dfrac{0.0591}{2} \log \dfrac{[0.0500 \text{ M}]}{[1.00 \text{ M}]} = 1.10 - (-0.04) = 1.14$ volts

    Answer: $\varepsilon = 1.14$ volts

    The concentration of the product, Zn^{2+}, is lower than the 1.0 M standard. The system is further from equilibrium, and the cell potential is *higher* than the standard voltage.

2. Calculate the voltage, ε, for the same cell when the cell has discharged to the point where the $[Cu^{2+}]$ is 0.010 M.

    Assume that the zinc electrode is not entirely consumed.

	Zn(s)	+ Cu^{2+}(aq)	⇒ Zn^{2+}(aq)	+ Cu(s)
Start	solid	1.000	0.0500	solid
Δ	−0.990	−0.990	+0.990	+0.990
Finish	Some	0.010	1.040	Some

    $\varepsilon = 1.10 - \dfrac{0.0591}{2} \log \dfrac{[1.040 \text{ M}]}{[0.010 \text{ M}]} = 1.10 - (+0.06) = 1.04$ volts

    Answer: $\varepsilon = 1.04$ volts

    The concentration of the product, Zn^{2+}, is higher than the 1.0 M standard. The concentration of the reactant, Cu^{2+}, is lower than the 1.0 M standard. The system is closer to equilibrium, and the cell potential is *lower* than the standard voltage.

Part II

Multiple-Choice Questions

About Section I
The Multiple-Choice Questions

Section I of the AP Chemistry exam constitutes 45% of the grade, and consists of 75-85 multiple choice questions. Usually 75 questions are asked and 90 minutes is allowed. The 85 question test was used *once*. It allows 115 minutes for the section and includes chemical equations. Your teacher will be able to tell you how many multiple-choice questions are on this year's test.

Chemical equations are reviewed where they are commonly tested—in Section II.

Only about 1 minute should be spent on each question. There are questions at the end of the section that will test concepts that are familiar, so it is important to get to them. Here are some additional hints.

1. The scoring formula is:

    Total Correct = # Right - (# Wrong ÷ 4).

    Answer only those questions where the right answer is known, or some of the choices can be eliminated. If 1 out of 5 items involving guesses is answered correctly, it's a wash. But 2 or more out of 5 answered correctly represents a significant gain.

2. Don't change an answer unless an obvious error is detected. Nine times out of 10 the first choice is the best one.

3. A time-tested axiom is "If you think long, you think wrong." The 'distracters'—wrong answers—are intentionally distracting.

4. Read the question carefully. Write on the exam paper. Underline the key chemical terms. Pay attention to dimensional units and significant figures.

5. Eliminate answers that are internally inconsistent. An exothermic reaction can *never* absorb energy. Sodium salts will *not* precipitate.

6. Some of the questions *are* surprisingly easy. Take the point—and the time—to move on to more challenging questions.

7. Depend on careful preparation and reading only! Forget the *rule* "When in doubt, pick answer (C) or (D)." The test is written by professionals. The key is most likely statistically checked. A string of two or three (A) answers is statistically possible, as is any other 'trend'.

Set 1

Atomic Structure and Periodicity

Directions for Questions 1-3:
Select an answer from the five atomic numbers given. You may use an atomic number once, more than once, or not at all.

(A) 4
(B) 8
(C) 9
(D) 17
(E) 85

1. This element has the highest ionization energy of the elements listed.

2. All isotopes of this element are naturally radioactive.

3. This element has two s- and only four p-electrons in its valence shell.

4. Which mass-spectrometer graph represents naturally occurring boron?

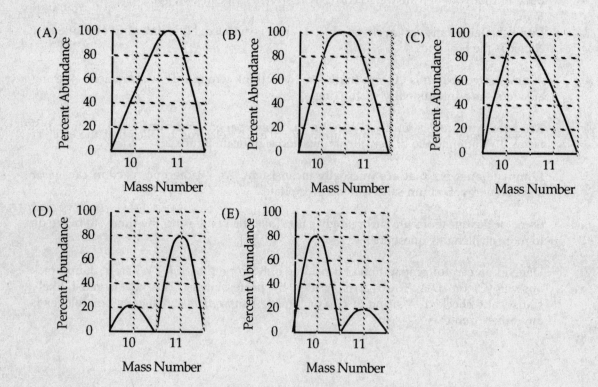

5. Two isotopes of silicon are Si-27 and Si-31. Both would be expected to have the same

(A) half-life
(B) nuclear charge
(C) atomic mass
(D) number of neutrons
(E) decay particle

6. Which of the electron transitions shown in the diagram would evolve the most energy?

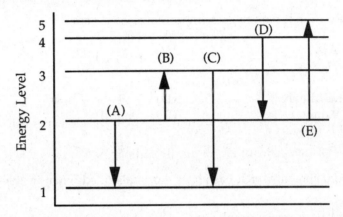

7. In a specific atom, how many electrons can there be with the quantum designation $n = 3$ and $l = 2$?

(A) 14
(B) 10
(C) 8
(D) 6
(E) 2

8. The assignment of the quantum number $l = 0$ to a $1s$-electron implies that the electron has no:

(A) energy
(B) orbital
(C) magnetic field
(D) angular momentum
(E) spin

9. Ions having the electron structure
$1s^2\ 2s^2 2p^6$
would be present in an aqueous solution of:

(A) LiBr
(B) $CaBr_2$
(C) NaF
(D) KCl
(E) SrI_2

10. How many *unpaired* electrons are in a nickel III ion, Ni^{3+}?

(A) 1
(B) 2
(C) 3
(D) 4
(E) 5

11. A magnesium atom (Mg) differs from a magnesium ion (Mg^{2+}) in that the *atom* has:

    (A) 3s electrons.
    (B) a higher ionization energy.
    (C) a greater stability.
    (D) a smaller radius.
    (E) fewer subshells.

12. Which would exhibit paramagnetism in the gaseous state?
    I   Al   IV Mg
    II Ar   V   Na
    III Hg

    (A) I, II and III only   (D) I, III and V only
    (B) I and V only   (E) Only II
    (C) II and IV only

13. The transition element which has all its *d*-orbitals filled when in the ground state is:

    (A) Ag   (D) Sn
    (B) Cr   (E) Ga
    (C) Fe

14. Which element would form colored cations and have multiple oxidation states?

    (A) Sc   (D) Zn
    (B) Mn   (E) Ag
    (C) F

15. Which set of elements is listed in order of increasing ionization energy?

    (A) Sb < As < P < S < Cl   (D) P < As < Sb < S < Cl
    (B) Cl < Sb < P < As < S   (E) Sb < As < Cl < S < P
    (C) As < Cl < P < S < Sb

16. Lithium (Li) has the highest ionization energy of the alkali metals (Group IA). But Li^+ has the lowest reduction potential of the group. Which is an explanation of this discrepancy?

    (A) Lithium ions have a high hydration energy.
    (B) Ionization energy is endothermic and reduction potential is exothermic.
    (C) Solid lithium has a strong metallic bond.
    (D) Less energy is required to remove an electron when the reaction is carried out in solution.
    (E) Ionization energy of metals and reduction potential of metal ions are measurements of exactly the same reaction, but going in different directions.

17. Why does gold have a higher first ionization energy than silver?

   (A) Gold is a noble metal, silver is not.
   (B) Gold causes the halogens to gain electrons more easily than silver.
   (C) Gold atoms are the same size as silver atoms, but have a higher nuclear charge.
   (D) Gold dissolves in a HCl-HNO3 mixture called 'aqua regia', silver only dissolves in HNO3.
   (E) The screening effect of f-subshell electrons is present in silver, but absent in gold.

18. Electron affinity generally increases as the atomic number increases in Period 2. However, the electron affinity *decreases* between

   I $_3$Li and $_4$Be   IV $_6$C and $_7$N
   II $_4$Be and $_5$B   V $_7$N and $_8$O
   III $_5$B and $_6$C

   (A) I only   (D) II and IV only
   (B) I and IV only   (E) V only
   (C) II and V only

19. In the decay series that starts with uranium-238, the first four steps consist of the emission of an alpha, two beta and then another alpha particle. What is the daughter isotope after the fourth emission?

   (A) $^{234}_{92}$U   (B) $^{230}_{90}$Th   (C) $^{234}_{90}$Th   (D) $^{230}_{88}$Ra   (E) $^{230}_{86}$Rn

20. Bismuth-210 has a half-life of 5.0 days. Approximately how many days would it take for 87.5% of a 1.00 mg sample of this isotope to beta-decay to polonium-210?

   (A) 1 day   (D) 9.4 days
   (B) 4.4 days   (E) 15 days
   (C) 5.8 days

Answer Key—Set 1

1.	(C)	6.	(C)	11.	(A)	16.	(A)
2.	(E)	7.	(B)	12.	(B)	17.	(C)
3.	(B)	8.	(D)	13.	(A)	18.	(B)
4.	(D)	9.	(C)	14.	(B)	19.	(B)
5.	(B)	10.	(C)	15.	(A)	20.	(E)

Explanations

1. **(C)** The highest ionization energies for elements are found in the top right corner of the periodic table. Only helium (He, element 2) and neon (Ne, element 10) have a higher first ionization energy than fluorine (F, element 9).

2. **(E)** All isotopes of elements whose atomic number is higher than 83 are naturally radioactive.

3. **(B)** The element has six valence electrons, and is in the fourth box of the p-block. It will be a member of the oxygen family of elements.

4. **(D)** Boron is found as two naturally occurring stable isotopes, boron-10 and boron-11. The atomic weight is a weighted average, and its value of 10.8 suggests there is 80% boron-11 and 20% boron-10.

5. **(B)** Since both isotopes are silicon, they would have to have the same atomic number and hence the same number of protons in the nucleus.

6. **(C)** Energy is evolved as the electron moves from higher to lower energy shells. The energy varies inversely with n^2, making the energy differences between shells greater at lower energies.

7. **(B)** The letter designation is the $3d$-subshell, which has 5 orbitals that can accommodate 10 electrons.

8. **(D)** The 1s-orbital is spherical in shape, and has no angular momentum.

9. **(C)** Sodium ions have 10 electrons. Ions generally have the same number of electrons as the nearest noble gas.

10. **(C)** Nickel has the electron structure $[Ar]3d^84s^2$. When three electrons are lost, the structure becomes $[Ar]3d^7$. Three of the five d-orbitals have unpaired electrons.

11. **(A)** Magnesium has the electron structure $[Ne]3s^2$. The $3s$-electrons are lost when a magnesium ion is formed.
    The magnesium atom, Mg, is larger and has more subshells than the ion. The atom also has a lower ionization energy and is therefore less stable than the ion (Mg^{2+} has a noble gas electron structure).

12. **(B)** Paramagnetism is caused by unpaired electrons. Aluminum and sodium have an odd number of electrons and one unpaired electron each.

13. **(A)** Silver has the electron structure $[Kr]4d^{10}5s^1$. Chromium and iron have incomplete d-subshells. Tin and gallium are not transition elements.

14. **(B)** Transition elements with incomplete d-subshells generally exhibit multiple oxidation states and have colored ions.

15. **(A)** The highest ionization energies occur in atoms at the top right corner of the periodic element. The lowest ionization energies occur in atoms at the bottom left hand corner of the periodic element.

16. **(A)** Reduction potentials are measured in 1.0 M water solutions. When an electron is removed from an atom to form an ion, energy is absorbed. When the ion hydrates (bond formation) energy is evolved. The large hydration energy associated with the small lithium ion lowers its reduction potential.

17. **(C)** Gold is the same approximate size as silver because of what is called the 'lanthanide contraction'. The $4f$-subshell lowers the shielding of the valence electrons from that which would be expected.

18. **(B)** Electron affinity is the amount of energy released when an electron is added to a neutral atom. Less energy is evolved when an electron is added to either an exactly filled or exactly half-filled sublevel, since these arrangements are stable electron-configurations.

19. **(B)** Two alphas (α is ^4_2He) reduce the mass of the daughter by 8 and the atomic number by 4. Two betas (β is $^0_{-1}e$)increase the atomic number of the daughter by 2. The net is to decrease the mass by 8 and the atomic number by 2.

20. **(E)** $\ln \dfrac{N_t}{N_o} = -\dfrac{0.693t}{t_{1/2}}$

    After time t, 87.5% (0.875) has decayed and 12.5% (0.125) remains.

    $\dfrac{N_t}{N_o}$ is the fraction *remaining*.

    $\ln 0.125 = -\dfrac{0.693t}{5.0 \text{ days}}$

    $t = \dfrac{(-2.08)(5.0 \text{ days})}{(-0.693)} = 15.0 \text{ days}$

Set 2
Chemical Bonding

For questions 1-4 select an answer from the five bond types given. You may use a bond type once, more than once, or not at all.

(A) Coordinate covalent bond
(B) Covalent bond
(C) Hydrogen bond
(D) Ionic bond
(E) Metallic bond

1. Which is the type of bond responsible for the linkage between atoms having a similar electron affinity and a high electronegativity?

2. Which is the bond which best accounts for the high melting points and low electrical conductivity of crystalline alkali metal salts?

3. Substances with this bonding have excellent thermal and electrical conductivity.

4. Transitional metal ions form d^2sp^3 octahedral complexes using this bond-type to link with ligands.

5. Which is always associated with chemical bonding?

   (A) Eight electrons in the valence shell of each atom in the compound.
   (B) Two electrons in the valence shell of each atom in the compound.
   (C) Filled valence shells in the atoms to be bonded.
   (D) The attractive forces in the compound are less than the attractive forces of the component atoms.
   (E) The attractive forces in the compound are greater than the attractive forces of the component atoms.

6. Which compound could be formed from the hypothetical elements M and Z? They have the Lewis electron-dot structures to the right.

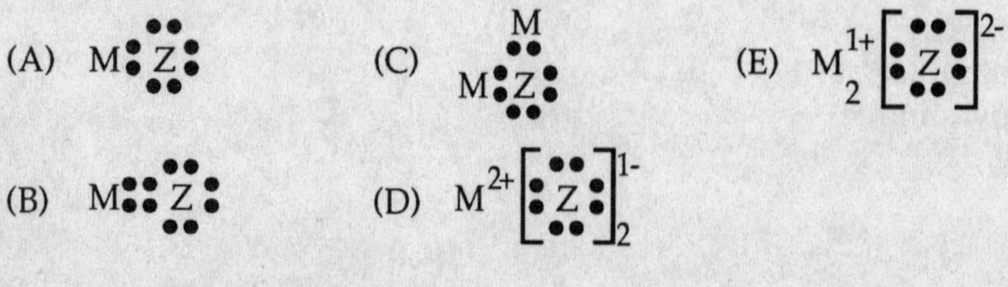

   (A) M:Z:
   (B) M::Z:
   (C) M:Z: with M above and below
   (D) $M^{2+}[:Z:]^{1-}_2$
   (E) $M^{1+}_2[:Z:]^{2-}$

7. Which is the valid Lewis structure for the sulfite ion (SO_3^{2-})?

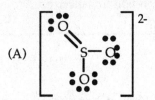 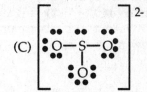 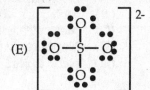

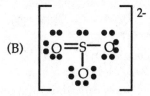 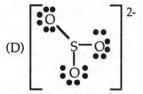

8. Which would have a resonance structure?

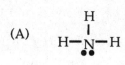 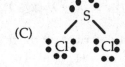 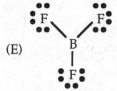

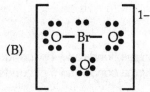 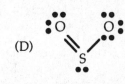

9. What is the geometric shape of the ClF_4^+ ion?

   (A) irregular pyramid (see-saw)
   (B) octahedron
   (C) square planar
   (D) tetrahedral
   (E) trigonal pyramidal

10. XF_3 has a trigonal pyramidal molecular structure. To which main group of the periodic table does 'X' belong?

    (A) 3 (B) 4 (C) 5 (D) 6 (E) 7

11. What is the geometry of BrF_3?

    (A) octahedron
    (B) planar triangle
    (C) trigonal bipyramidal
    (D) trigonal pyramidal
    (E) T-shaped

12. In the PCl_6^- ion, what is the Cl-P-Cl bond angle?

    (A) 60° (B) 90° (C) 109.5° (D) 120° (E) 180°

13. Molecular orbital theory predicts that Ne$_2$ will not form bonds because:
    (A) Neon has equal numbers of bonding electrons and anti-bonding electrons.
    (B) Molecules with electrons in σ* orbitals are unstable.
    (C) Molecules with electrons in π* orbitals are unstable.
    (D) Ne$_2$ can only form weak π$_{2p}$ bonds.
    (E) Neon is paramagnetic.

14. According to molecular orbital theory, which has the highest bond energy and the smallest bond distance?
    (A) Be$_2$ (B) NO   (C) NO$^+$ (D) NO$^-$ (E) O$_2$

15. Which is the molecular orbital structure for an oxygen molecule, O$_2$?
    (A) $(\sigma_{2s}^b)^2 (\sigma_{2s}^*)^2 (\pi_{2p}^b)^4 (\pi_{2p}^*)^4$
    (B) $(\sigma_{2s}^b)^2 (\sigma_{2s}^*)^2 (\sigma_{2p}^b)^2 (\sigma_{2p}^*)^2 (\pi_{2p}^b)^4$
    (C) $(\sigma_{2s}^b)^2 (\sigma_{2s}^*)^2 (\sigma_{2p}^b)^2 (\pi_{2p}^b)^3 (\pi_{2p}^*)^3$
    (D) $(\sigma_{2s}^b)^2 (\sigma_{2s}^*)^2 (\sigma_{2p}^b)^2 (\pi_{2p}^b)^4 (\pi_{2p}^*)^2$
    (E) $(\sigma_{2s}^b)^2 (\sigma_{2s}^*)^2 (\sigma_{2p}^b)^2 (\pi_{2p}^b)^4 (\pi_{2p}^*)^4 (\sigma_{2p}^*)^2$

16. Which description best describes all the bonds in acrylonitrile, CH$_2$CHCN?

    H₂C=CH−C≡N

    (A) 9 pi bonds
    (B) 9 sigma bonds
    (C) 4 sigma bonds and 5 pi bonds
    (D) 5 sigma bonds and 4 pi bonds
    (E) 6 sigma bonds and 3 pi bonds

17. Which molecule has sp^2 hybridization of a carbon atom?
    (A) CH$_2$CH$_2$
    (B) CH$_3$CH$_3$
    (C) CHCH
    (D) CH$_4$
    (E) CO$_2$

18. What is the correct name of K$_2$[PtCl$_4$]?
    (A) potassium tetrachloroplatinum
    (B) potassium tetrachloroplatinum II
    (C) potassium tetrachloroplatinum IV
    (D) potassium tetrachloroplatinate II
    (E) potassium tetrachloroplatinate IV

19. What is the number of *ions* in the coordination compound, [Ir(NH$_3$)$_3$Cl$_3$]Cl$_3$?
    (A) two   (B) four   (C) six   (D) seven   (E) ten

20. Which could *not* be a ligand?
    (A) H$_2$O
    (B) CO
    (C) CN$^-$
    (D) NH$_2$CH$_2$=CH$_2$NH$_2$
    (E) CH$_4$

Chemical Bonding

Answer Key—Set 2

1.	(B)	6.	(D)	11.	(E)	16.	(E)
2.	(D)	7.	(C)	12.	(B)	17.	(A)
3.	(E)	8.	(D)	13.	(A)	18.	(D)
4.	(A)	9.	(A)	14.	(C)	19.	(B)
5.	(E)	10.	(C)	15.	(D)	20.	(E)

Explanations

1. **(B)** Covalent bonds occur between nonmetals, which are characterized by high electron affinities and high ionization energies.

2. **(D)** The alkali metal salts are made up of cations and anions electrostatically attracted to each other forming a strong crystalline structure. Since the ions cannot move, the salt will not conduct electricity in the solid state.

3. **(E)** Metals are characterized by delocalized electrons. These mobile electrons account for the high thermal and electrical conductivity.

4. **(A)** The transition metal ions are electron-deficient, which allows the acceptance by the central atom of nonbonded (lone-pair) electrons from ligands.

5. **(E)** The attractive forces in compounds must be greater than those in the component atoms or the bond would not occur. Most, but not all atoms bond to form stable octets. Only hydrogen has two electrons in the valence shell after bonding. Filled valence shells do not tend to bond (e.g. noble gases).

6. **(D)** Each 'M' will donate or share 2 electrons. Each 'Z' will accept one electron or share a pair of electrons with each of the two 'M' atoms.

7. **(C)** Count the electrons. The sulfur and the three oxygen atoms have a total of 24 valence electrons. The 2$^-$ charge on the ion accounts for two more, for a total of 26. Three bonds use 6, leaving 20 electrons (10 pairs) to complete octets around each atom.

8. **(D)** Resonance occurs in structures with single *and* double bonds. Equivalent Lewis structures for SO_2 can be written by switching the single and double bonds.

9. **(A)** ClF_4^+ has 10 electrons surrounding the chlorine and a coordination number of 5. The nonbonding pair (lone-pair electrons) will be in the plane of the trigonal bipyramid.

10. **(C)** The Group V elements have 5 valence electrons. The result is a structure with a lone pair and 3 unpaired (bonding) electrons.

11. **(E)** BrF_3 has 10 electrons surrounding the bromine and a coordination number of 5. Two nonbonding pairs will be on the plane of the trigonal pyramid, resulting in a T-shaped bonding structure.

12. **(B)** The PCl_6^- ion has 12 bonding electrons or 6 bonds between the chlorine atoms and the central phosphorus atom, resulting in an octahedral shape.

13. **(A)** Ne$_2$, with a total of 16 valence electrons, would have 8 in bonding orbitals and 8 in antibonding orbitals. The bond order is zero.

14. **(C)** NO$^+$ has a bond order of 3. NO has a bond order of 2½. O$_2$ and NO$^-$ each have a bond order of 2. Be$_2$ has a bond order of zero. The high bond orders exhibit high bond energy and small bond lengths.

15. **(D)** Lower energy bonding orbitals are filled before higher energy antibonding orbitals.

16. **(E)** Single bonds are sigma bonds. In a multiple bond, the one bond is a sigma (σ) bond and other bonds are pi (π) bonds.

17. **(A)** sp^2 hybridization occurs. The carbon atom forms three bonds in this case—two single bonds and one double covalent bond.

18. **(D)** When the complex ion is the anion, the ending 'ate' is used. The oxidation number of the two potassium ions is +2, and that of the four chloro ligands is –4. The oxidation state of the platinum ion must be +2 to give a neutral salt.

19. **(B)** The complex ion is stable and remains together. There are three chloride anions in this salt and one complex anion.

20. **(E)** There are no lone–pair electrons in CH$_4$.

Set 3

Stoichiometry

1. Which sample of gas contains the same number of molecules as 68.0 g of hydrogen sulfide (H_2S)?

    (A) 4.0 gram of hydrogen gas (H_2).
    (B) 25.5 grams of ammonia gas (NH_3).
    (C) 32.0 gram of oxygen gas (O_2).
    (D) 66.0 grams of carbon dioxide gas (CO_2).
    (E) 352 grams of uranium hexafluoride gas (UF_6).

2. Which 1.0 gram sample contains the greatest number of atoms?

    (A) Aluminum chloride ($AlCl_3$)
    (B) Ammonia (NH_3)
    (C) Fluorine (F_2)
    (D) Neon (Ne)
    (E) Propane (C_3H_8)

3. Element M reacts with oxygen to produce a pure sample of MO_2. Find the molar mass and the identity of M if 9.6 g of oxygen reacts with 16.5 g of M to produce 26.1 g of MO_2.

    (A) 12, carbon
    (B) 14, nitrogen
    (C) 32, sulfur
    (D) 28, silicon
    (E) 55, manganese

4. Elements X and fluorine react to form two different compounds. In the first reaction 0.480 g of X reacts with 1.000 g of F. In the second 0.960 g of X reacts with 1.000 g of F. Which are possible formulas for the two compounds?

	First Compound	Second Compound
(A)	XF	X_2F_2
(B)	XF_2	XF
(C)	XF_2	X_2F
(D)	X_2F	XF_2
(E)	X_2F_3	X_3F

5. Analysis of an organic compound revealed it contains only 0.566 g hydrogen, 2.641 g nitrogen, and 6.793 g carbon. What is the empirical formula of the compound?

    (A) C_3H_3N
    (B) C_6H_6N
    (C) $C_{12}HN_5$
    (D) $CH_3CH_2NH_2$
    (E) HCN

6. A compound has the empirical formula C_3H_4O and a molar mass of 168.2 g mol^{-1}. Its molecular formula is:

   (A) C_3H_4O
   (B) $C_9H_{14}NO_2$
   (C) $C_{12}H_{16}O$
   (D) $C_9H_{12}O_3$
   (E) $C_8H_8O_4$

7. When 1.24 g of an organic compound with the formula $C_xH_yO_z$ is burned in excess oxygen, 1.76 g of carbon dioxide (CO_2) and 1.08 g of water vapor (H_2O) are obtained. What is the empirical formula of the compound?

   (A) CHO
   (B) CH_2O
   (C) CH_3O
   (D) $C_2H_3O_4$
   (E) C_3H_5O

8. When germanium (Ge) is reacted with an excess of oxygen, germanium oxide is formed. If the reaction vessel plus the GeO_2 formed in this reaction weighs 43.78 g after the reaction, and it has gained 8.90 g during the reaction, what is the weight of the empty reaction vessel?

   (A) 8.90 g
   (B) 14.69 g
   (C) 23.60 g
   (D) 30.96 g
   (E) 34.88 g

9. Octane, (C_8H_{18}, MM=114.08) is burned completely in an excess of air to give carbon dioxide, (CO_2), and water vapor (H_2O). How many grams of carbon dioxide can be produced by the complete combustion of 28.52 g of octane?

   (A) 11.0 g
   (B) 22.0 g
   (C) 44.0 g
   (D) 88.0 g
   (E) 176.0 g

10. When 0.500 mol of $AlCl_3$(aq) is mixed with 0.300 mol of K_2CrO_4(aq), what is the maximum number of moles of $Al_2(CrO_4)_3$(s) that can be formed?

    (A) 0.100 mol
    (B) 0.250 mol
    (C) 0.500 mol
    (D) 0.600 mol
    (E) 0.900 mol

11. How many milliliters of 6.0 M hydrochloric acid are required to prepare 250. mL of a 1.20 M HCl(aq) solution?

    (A) 200. mL
    (B) 125. mL
    (C) 50.0 mL
    (D) 28.8 mL
    (E) 12.5 mL

12. The density of a solution of 20.0 % (by weight) nitric acid, HNO_3 solution (MM = 63.0), in water is 1.117 g/mL (1.117 kg/L).
What is the molarity of this solution?

(A) 88.8 M
(B) 3.54 M
(C) 3.17 M
(D) 2.84 M
(E) 1.13 M

13. A 12.0% solution of KOH(aq) (MM = 56.11) has a density of 1.1079 g/mL. Which set of answers is the mole fraction, molality, and molarity, respectively, for this solution?

(A) 0.042, 2.37, 2.43
(B) 0.042, 2.43, 2.37
(C) 2.43, 0.042, 2.37
(D) 2.43, 2.37, 0.042
(E) 2.37, 2.43, 0.042

14. To determine the molar concentration of a 20.0 mL sulfuric acid, H_2SO_4, solution, aqueous $Ba(OH)_2$ is added in excess, assuring complete reaction. The resulting pure precipitate, $BaSO_4$ (MM = 233.4), weighs 0.560 g. What is the initial concentration of the sulfuric acid?

(A) 6.54 M
(B) 0.480 M
(C) 0.240 M
(D) 0.120 M
(E) 0.060 M

15. When 30.0 mL of 0.100 M iron (III) nitrate solution, $(Fe(NO_3)_3)$ is reacted with a 0.200 M potassium oxalate solution, $(K_2C_2O_4)$, iron (III) oxalate is precipitated. What is the *minimum* volume of $K_2C_2O_4$(aq) required for a maximum yield of $Fe_2(C_2O_4)_3$ precipitate?

(A) 15.0 mL
(B) 20.0 mL
(C) 22.5 mL
(D) 40.0 mL
(E) 60.0 mL

16. What is the molarity of chloride ion, Cl^-(aq), in 100. mL of a 0.500 M barium chloride, $BaCl_2$, solution?

(A) 0.050 M
(B) 0.100 M
(C) 0.250 M
(D) 0.500 M
(E) 1.00 M

17. The molar mass is determined by dissolving 5.00 g of a compound in 50.0 g of benzene. The freezing point was lowered by 2.5°C. The freezing point constant, K_f, for benzene is 5.0°C $molal^{-1}$. What is the molar mass of the compound?
(A) 25
(B) 50.
(C) 100.
(D) 150.
(E) 200.

18. Methanol, CH₃OH, is more effective than an equal weight of ethanol, CH₃CH₂OH, in lowering the freezing point of 100. g of water because the methanol

    (A) is more soluble in water.
    (B) has a lower boiling point.
    (C) forms stronger hydrogen bonds.
    (D) has more dissolved particles.
    (E) has a higher vapor pressure.

19. A solution of a nonvolatile, nonionizing solute in toluene is prepared. Which is true?

	Vapor pressure	Melting point	Boiling point
(A)	Increased	Increased	Decreased
(B)	Decreased	Decreased	Increased
(C)	Increased	Increased	Increased
(D)	Decreased	Increased	Increased
(E)	Decreased	Decreased	Decreased

20. What will be the vapor pressure at 25°C of a 30.0% by mass solution of the nonvolatile solute glucose, $C_6H_{12}O_6$ (MM = 180), in water? The vapor pressure of water at 25°C is 23.8 mmHg.

    (A) 7.1 mmHg
    (B) 10.2 mmHg
    (C) 16.7 mmHg
    (D) 22.8 mmHg
    (E) 23.3 mmHg

Answer Key—Set 3

1.	(A)	6.	(D)	11.	(C)	16.	(E)		
2.	(E)	7.	(C)	12.	(B)	17.	(E)		
3.	(E)	8.	(B)	13.	(B)	18.	(D)		
4.	(B)	9.	(D)	14.	(D)	19.	(B)		
5.	(A)	10.	(A)	15.	(C)	20.	(D)		

Explanations

1. **(A)** There are 2.00 moles of H_2S (MM = 34.0) in 68.0 grams. There are 2.0 moles of H_2 molecules (MM = 2.0) in 4.0 grams.

2. **(E)** The number of moles is 1.0 g divided by the molar mass and multiplied by the number of atoms in the formula. Propane has $((1.0 \div 44.0) \times 11) = 0.25$ moles of atoms.

3. **(E)** $n_{oxygen} = 0.60 \text{ mol}$

   $n_M = \dfrac{1 \text{ mol M}}{2 \text{ mol O}} \times 0.60 \text{ mol O} = 0.30 \text{ mol}$

   The atomic weight of M = $\dfrac{16.5 \text{ g}}{0.30 \text{ mole}} = 55 \text{ g mol}^{-1}$

4. **(B)** The weights suggest that the second compound has twice as many atoms of X per atom of fluorine as the first. XF has half as many atoms of F per atom of X as XF_2. This is another way to say the same thing.

5. **(A)**

Element	Weight	Moles	Ratio
Carbon	6.793 g	0.566	3.00
Hydrogen	0.566 g	0.566	3.00
Nitrogen	2.641 g	0.189	1.00

6. **(D)** C_3H_4O has a molar mass of 56.03. The multiple is 3 ($168.2 \div 56.03 = 3.005$). $(C_3H_4O)_3 = C_9H_{12}O_3$

7. **(C)**

Element	Mol	Weight	Mol	Ratio
Carbon	0.040	0.48 g	0.040	1.0
Hydrogen	0.120	0.12 g	0.120	3.0
Oxygen	NA	0.64 g	0.040	1.0
Total		1.24 g		

8. **(B)**

	Ge	+	O_2	$\Rightarrow$	GeO_2
Weight	20.18 g		8.90 g		29.09 g
Mol	0.278		0.278		0.278

Wt_{vessel} = 43.78 g − 29.09 g = **14.69 g**

9. **(D)** $C_8H_{18} + 12\frac{1}{2} O_2 \Rightarrow 8CO_2 + 9H_2O$

$n_{C_8H_{18}}$ = 28.52 g ÷ 114.08 g mol^{-1} = 0.2500

n_{CO_2} = $\dfrac{8 \text{ mol } CO_2}{1 \text{ mol } C_8H_{18}} \times n_{octane}$

n_{CO_2} = 8 × 0.2500 = 2.000 mol

wt_{CO_2} = 2.000 × 44.0 = **88.0 g**

10. **(A)** $2Al^{3+} + 3CrO_4^{2-} \Rightarrow Al_2(CrO_4)_3$

$n_{Al_2(CrO_4)_3}$ = ½ $n_{Al^{3+}}$ = ½ × 0.500 = 0.250 mol

$n_{Al_2(CrO_4)_3}$ = ⅓ $n_{CrO_4^{2-}}$ = ⅓ × 0.300 = 0.100 mol

The chromate ion is the limiting reactant, and 0.100 mol is the maximum quantity of aluminum chromate formed by the complete reaction of 0.300 moles of CrO_4^{2-}.

11. **(C)** The dilution equation is: $M_1V_1 = M_2V_2$

V_{HCl} = $\dfrac{0.250 \text{ L} \times 1.20 \text{ mol L}^{-1}}{6.0 \text{ mol L}^{-1}}$ = 0.050 L = **50. mL**

12. **(B)** Basis: 1000. $g_{solution}$ = 1.000 $kg_{solution}$

n_{HNO_3} = 200. g ÷ 63.0 g mol^{-1} = 3.17 mol

$L_{solution}$ = 1.000 $kg_{solution}$ ÷ 1.117 kg L^{-1} = 0.895 L

Molarity = 3.17 mol ÷ 0.895 L = **3.54 M**

13. **(B)** Basis: 1000. $g_{solution}$ = 1.000 $kg_{solution}$

n_{KOH} = 120. g ÷ 56.0 g mol^{-1} = 2.14 mol

n_{HOH} = 880. g ÷ 18.0 g mol^{-1} = 48.8 mol

x = 2.14 ÷ (2.14 + 48.8) = 0.042

Molality = $\dfrac{2.14 \text{ mol KOH}}{0.880 \text{ kg HOH}}$ = 2.43 m

$L_{solution}$ = $\dfrac{1000 \text{ kg}_{solution}}{1.1079 \text{ kg L}^{-1}}$ = 0.903 L

Molarity = 2.14 mol ÷ 0.903 L = **2.37 M**

Note that:
molarity (M) is less than molality (m)
mole fraction (x) is much less than molarity (M) and molality (m)

14. **(D)** $Ba(OH)_2(aq) + H_2SO_4(aq) \Rightarrow BaSO_4(s) + 2H_2O(l)$

n_{BaSO_4} = 0.560 g ÷ 233.4 g mol^{-1}

n_{BaSO_4} = 0.00240 mol

$n_{H_2SO_4}$ = n_{BaSO_4}

Molarity = 0.00240 mol ÷ 0.0200 L = **0.120 M**

15. **(C)** $2Fe(NO_3)_3(aq) + 3K_2C_2O_4(aq) \Rightarrow Fe_2(C_2O_4)_3(s) + 6KNO_3(aq)$

$n_{Fe(NO_3)_3}$ = 0.0300 L × 0.100 mol L^{-1} = 0.00300 mol

$n_{K_2C_2O_4}$ = $\dfrac{3 \text{ mol } K_2C_2O_4}{2 \text{ mol } Fe(NO_3)_3} \times n_{Fe(NO_3)_3}$

$n_{K_2C_2O_4}$ = 3/2 × 0.00300 = 0.00450 mol

$V_{K_2C_2O_4}$ = $\dfrac{0.00450 \text{ mol}}{0.20 \text{ mol L}^{-1}}$ = 0.0225 L

$V_{K_2C_2O_4}$ = **22.5 mL**

16. **(E)** $BaCl_2(aq) \Rightarrow Ba^{2+}(aq) + 2Cl^{1-}(aq)$

n_{Cl^-} = 2 × n_{BaCl_2}

The same volume will contain twice the moles of chloride ion.

Molarity = 2 × 0.500 M = **1.00 M**

17. **(E)**

Weight = 5.00 g$_{solute}$ ÷ 0.050 kg$_{benzene}$ = 100. g kg^{-1}

molality = $\Delta T \div K_f$ = 2.5°C ÷ 5.0°C m^{-1} = 0.50 mol kg^{-1}

Molar mass = $\dfrac{100. \text{ g kg}^{-1}}{0.50 \text{ mol kg}^{-1}}$ = **200. g mol^{-1}**

18. **(D)** The same weight of methanol will contain more moles of molecules because of its *lower* molar mass.

19. **(B)** A solution has a lower vapor pressure than the solvent due to an increase in attraction between molecules. This lower vapor pressure results in a *lower* freezing (melting) temperature and a *higher* boiling temperature than the pure solvent.

20. **(D)** Basis: 100. g of solution

n_{solute} = $\dfrac{30.0 \text{ g}}{180 \text{ g mol}^{-1}}$ = 0.167 mol

$n_{solvent}$ = $\dfrac{70.0 \text{ g}}{18.0 \text{ g mol}^{-1}}$ = 3.89 mol

$x_{solvent}$ = $\dfrac{3.89 \text{ mol}}{(3.89 \text{ mol} + 0.17 \text{ mol})}$ = 0.958

$P_{solvent}$ = $x_{solvent} P^°_{solvent}$ = 0.958 × 23.8 mmHg = **22.8 mmHg**

Set 4

States of Matter

1. Which is a possible explanation why silicon carbide, SiC, is twice as hard as zinc oxide, ZnO, even though they both have a hexagonal crystal lattice? Silicon carbide:

    (A) has atoms which fit into the lattice more neatly.
    (B) has higher van der Waals forces.
    (C) has higher electrostatic forces.
    (D) is a covalent network solid.
    (E) is a stronger dipole.

2. The spontaneous endothermic solution of a salt in water suggests that:

    (A) all salts release energy when they dissolve.
    (B) all salts absorb energy when they dissolve.
    (C) the bond energy is more than the hydration energy.
    (D) the bond energy is less than the hydration energy.
    (E) the bond energy is equal to the hydration energy.

3. Molecular liquids:

    (A) boil at low temperatures.
    (B) generally have high vapor pressures.
    (C) dissolve readily in polar solvents.
    (D) have covalent bonds between the molecules.
    (E) conduct an electric current.

4. Which explains the trend in the melting temperatures of the four molecules?
    (A) CI_4 is polar.
    (B) CF_4 is ionic.
    (C) CF_4 is hydrogen bonded.
    (D) CI_4 is a network solid.
    (E) CI_4 has more electrons.

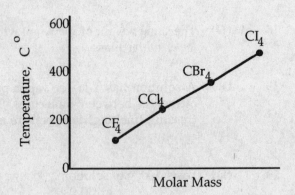

5. During the heating of a solid *at its melting temperature* the **average**:

    (A) kinetic energy is constant, but the average potential energy increases.
    (B) potential energy is constant, but the average kinetic energy increases.
    (C) potential and kinetic energies remain the same.
    (D) kinetic energy of the liquid phase is greater than the solid phase.
    (E) potential energy of the solid phase is greater than the liquid phase.

Questions 6 and 7 refer to the diagram.

6. What is the point that represents the temperature above which the gas will not condense at a high pressure?

   (A) A
   (B) B
   (C) C
   (D) D
   (E) E

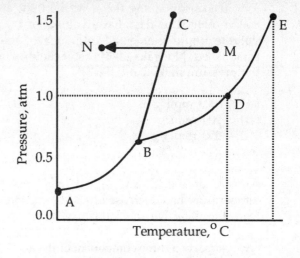

7. A phase change that occurs from point M to point N is:
   (A) condensation.
   (B) crystallization.
   (C) boiling.
   (D) sublimation.
   (E) vaporization.

8. The conditions of a confined gas are changed. It is observed that the pressure is higher, but the molecules hit the walls of the container with the original force. Which is true?
   (A) The temperature is the same, but the volume is larger.
   (B) The temperature and the volume are the same.
   (C) The temperature is higher, but the volume is the same.
   (D) The temperature is the same, but the volume is smaller.
   (E) The temperature and the volume are both smaller.

9. The temperature of a sample of CO_2 gas is increased. Which can be true?

	Volume	Pressure	Density
(A)	Constant	Increased	Increased
(B)	Constant	Increased	Constant
(C)	Constant	Constant	Decreased
(D)	Increased	Increased	Constant
(E)	Increased	Increased	Increased

10. A correction must be made for intermolecular attractions at low temperatures and/or high pressures. Using 'a' as this correction factor, choose the equation that, for one mole of a gas, corrects for intermolecular attractions.

    (A) $\left(P + \left(\dfrac{a}{V^2}\right)\right)V = RT$

    (B) $\left(P - \left(\dfrac{a}{V^2}\right)\right)V = RT$

    (C) $P(V - a) = RT$

    (D) $P(V + a) = RT$

    (E) $PV = RT$

122 Multiple-Choice Questions

11. Two flasks containing the same inert gas are at the same temperature and pressure of 800. mmHg. One flask has a volume of 1.0 L and the other a volume of 2.0 L. Enough volatile liquid is injected into each of the flasks to allow phase equilibrium to be established. No leakage occurs. If the pressure in the 1.0 L flask is 900 mmHg, what is the pressure in the 2.0 L flask?

    (A) 1000. mmHg
    (B) 900. mmHg
    (C) 850. mmHg
    (D) 800. mmHg
    (E) 450. mmHg

12. A solid at 100.°C was sealed in a vessel containing dry air. After one hour the temperature had decreased to 80°C and the pressure had increased. It can be concluded from the data that the solid may have:

    (A) adsorbed some component of the air.
    (B) reacted with the oxygen in the air.
    (C) sublimed.
    (D) boiled.
    (E) melted.

13. A gas has a density of 2.68 g L^{-1} at a pressure of 1.50 atm and a temperature of 27°C. What is its molar mass?

    (A) 44 g mol^{-1}
    (B) 59 g mol^{-1}
    (C) 66 g mol^{-1}
    (D) 89 g mol^{-1}
    (E) 109 g mol^{-1}

14. One gram (1.00 g) of a gaseous hydrocarbon occupies 0.821 L at 1.00 atm and 147°C. The compound is:

    (A) CH_4
    (B) C_2H
    (C) C_2H_4
    (D) C_3H_6
    (E) C_4H_{82}

15. A 1.00 L container at 273°C contains 6.00 moles of helium gas. What is the pressure of the gas?
    (A) 7.5 atm
    (B) 22.4 atm
    (C) 89.6 atm
    (D) 134 atm
    (E) 269 atm

16. A mixture of nitrogen, N_2, and oxygen, O_2, occupies a 1.0 L volume at a pressure of 800. mmHg. There are 4 times as many nitrogen molecules as oxygen molecules.
    All the N_2 molecules are removed and placed in a different 1.0 L container at the same temperature. What would be the pressure in the new container, assuming ideal behavior?

    (A) 800. mmHg (D) 200. mmHg
    (B) 640. mmHg (E) 160. mmHg
    (C) 600. mmHg

17. Which is necessary to determine the vapor pressure of water in a gas which has been collected by the displacement of water in a gas measuring tube (eudiometer)?

    (A) The volume of the gas (D) The water solubility of the gas
    (B) The barometric pressure (E) The temperature of the water
    (C) The volume of the water

18. The vapor pressure of hexane, C_6H_{14}, is 100. mmHg at 16°C. What is the vapor pressure at 16°C of a solution containing 0.500 mol of nonvolatile solute dissolved in 4.50 mol of heptane?

    (A) 180. mmHg (D) 90.0 mmHg
    (B) 110. mmHg (E) 10.0 mmHg
    (C) 100. mmHg

19. Beaker 'A' contains distilled water and Beaker 'B' contains an equal volume of a 1.0 molar solution of sugar in water. They are placed side-by-side in a sealed small aquarium. Which will be true one day later?

    (A) Beaker 'B' will contain more liquid than Beaker 'A'.
    (B) Beaker 'A' will contain more liquid than Beaker 'B'.
    (C) The sugar will be less than 1.0 m in Beaker 'B'.
    (D) The sugar will be more than 0.0 m in Beaker 'A'.
    (E) Both beakers contents will be the same as when they started.

20. Methane, CH_4, effused through an small hole at a rate of 5.00 mL sec^{-1}. What would be the rate of effusion of sulfur trioxide, SO_3, at the same temperature and through the same hole?

    (A) 25.0 mL sec^{-1} (D) 2.24 mL sec^{-1}
    (B) 11.2 mL sec^{-1} (E) 1.00 mL sec^{-1}
    (C) 5.00 mL sec^{-1}

Answer Key—Set 4

1.	(D)	6.	(E)	11.	(B)	16.	(B)
2.	(C)	7.	(B)	12.	(C)	17.	(E)
3.	(A)	8.	(D)	13.	(A)	18.	(D)
4.	(E)	9.	(B)	14.	(D)	19.	(A)
5.	(A)	10.	(A)	15.	(E)	20.	(B)

Explanations

1. **(D)** The elements of Group 4A sometimes form a three-dimensional network covalency. Silicon carbide, SiC, is used in the manufacture of grinding wheels and sandpaper.

2. **(C)** When a salt dissolves in water, energy is absorbed as the crystal lattice is broken, and energy is released as the ions hydrate to the water molecules. The energy of the bond breaking exceeds the energy of the bond formation for an endothermic reaction (net energy absorbed).

3. **(A)** Molecular liquids have low boiling points. The molecules in molecular liquids are attracted to each other by van der Waals' forces. These forces are hydrogen bonding, are dipoles attractions or London dispersion forces between nondipoles. These are weak attractions when compared to metallic, ionic and network covalent bonding.

4. **(E)** The four molecules are nonpolar and are held together by London dispersion forces. The forces are largest for the CI_4 molecule which has the highest number of electrons of the group.

5. **(A)** The temperature is constant at the melting temperature and therefore the average kinetic energy will not change. The average potential energy will increase due to the increased average distance between molecules in the liquid phase.

6. **(E)** Only the gaseous phase can exist above the critical temperature.

7. **(B)** The line BC represents the melting temperature (or fusion) curve.

8. **(D)** The molecules hitting the wall with the same force indicates velocity, the temperature and the kinetic energy are the same. Pressure and volume are inversely proportional. The volume must have been decreased to cause the increase in pressure.

9. **(B)** The increase in temperature will cause an increase in the pressure if the volume is constant. The density will be constant if the weight and volume are the same.

10. **(A)** The attraction between molecules will cause a decrease in the pressure. The correction factor must be added to the pressure.

11. **(B)** The vapor pressure of a liquid is dependent only on temperature.

12. **(C)** The sublimation of a solid to the gaseous phase both lowers the temperature (the breaking apart of the molecules absorbs energy—an endothermic reaction) and increases the pressure (there is an increase in the number of collisions with the walls of the container).

13. **(A)** $MM = \dfrac{DRT}{P} = \dfrac{2.68\ g\ L^{-1} \times 0.0821\ L\ atm\ mol^{-1}\ K^{-1} \times 300\ K}{1.50\ atm}$

    Molar mass = **44.0 g mol^{-1}**

14. **(D)** $MM = \dfrac{gRT}{PV} = \dfrac{1.00\ g \times 0.0821\ L\ atm\ mol^{-1}\ K^{-1} \times 420\ K}{1.00\ atm \times 0.821\ L}$

    Molar mass = **42.0 g mol^{-1}**, matching C$_3$H$_6$ (MM = 42.0)

15. **(E)** $P = \dfrac{nRT}{V} = \dfrac{6.00\ mol \times 0.0821\ L\ atm\ mol^{-1}\ K^{-1} \times 546\ K}{1.00\ L}$

    Pressure = **269.0 atm**

16. **(B)** The nitrogen is $^4/_5$ of the gaseous mixture. Its partial pressure will be the same fraction of the total 800. mmHg pressure.

    P$_{nitrogen}$ = $^4/_5 \times$ 800. mmHg = **640. mmHg**

17. **(E)** The vapor pressure of a liquid is dependent only on temperature.

18. **(D)** Raoult's Law predicts the vapor pressure of the solution.

    $x_{solvent}$ = $\dfrac{4.50\ mol}{(0.50 + 4.50)}$ = 0.900 mol

    P$_{solvent}$ = $x_{solvent}$ P°$_{solvent}$

    P$_{solvent}$ = 0.900 × 100. mmHg = **90.0 mmHg**

19. **(A)** Aqueous solutions have lower vapor pressures than pure water. The attraction between the solute and the solvent is higher than between water molecules. The sugar solution and the pure water will evaporate until vapor-liquid equilibrium is established at the temperature of the aquarium. More of the pure water will evaporate due to its higher vapor pressure at the same temperature.

20. **(B)** Graham's law predicts the effusion rate.

    $\dfrac{\text{Rate of SO}_3}{\text{Rate of CH}_4} = \sqrt{\dfrac{MM\ CH_4}{MM\ SO_3}}$

    Rate of SO$_3$ = $\sqrt{\dfrac{16.0}{80.0}} \times 5.00$ ml sec^{-1}

    Rate of SO$_3$ = **2.24 mL sec^{-1}**

Set 5

Reaction Kinetics

1. The reaction
   $$2NO(g) + Br_2(g) \Rightarrow 2NOBr(g)$$
   is second order with respect to NO and first order with respect to Br_2. Which is a correct rate law equation for this reaction?

   (A) $+\dfrac{\Delta[NO]}{\Delta t} = +\dfrac{\Delta[Br_2]}{\Delta t} = k[NO]^2[Br_2]$

   (B) $-\dfrac{1}{2}\dfrac{\Delta[NOBr]}{\Delta t} = +\dfrac{\Delta[Br_2]}{\Delta t} = k[NO]^2[Br_2]$

   (C) $+\dfrac{1}{2}\dfrac{\Delta[NOBr]}{\Delta t} = -\dfrac{1}{2}\dfrac{\Delta[NO]}{\Delta t} = k[NOBr]^2$

   (D) $-\dfrac{\Delta[NO]}{\Delta t} = -\dfrac{1}{2}\dfrac{\Delta[Br_2]}{\Delta t} = k[NOBr]^2$

   (E) $-\dfrac{1}{2}\dfrac{\Delta[NO]}{\Delta t} = -\dfrac{\Delta[Br_2]}{\Delta t} = k[NO]^2[Br_2]$

2. For the reaction:
   $$2O_3(g) \Rightarrow 3O_2(g)$$
   the reaction rate of O_3 is -2.0 mmHg s^{-1}.
   What is the reaction rate of the O_2 during the same time period?

   (A) 3.0 mmHg s^{-1}
   (B) 2.0 mmHg s^{-1}
   (C) 1.5 mmHg s^{-1}
   (D) 1.3 mmHg s^{-1}
   (E) 0.67 mmHg s^{-1}

Questions 3-7 refer to these data for the reaction:
$$2NO(g) + 2H_2(g) \Rightarrow N_2(g) + 2H_2O(g)$$

Run	NO Pressure	H$_2$ Pressure	Rate
1	0.375 atm	0.500 atm	6.43 x 10^{-4} atm s^{-1}
2	0.375 atm	0.250 atm	3.15 x 10^{-4} atm s^{-1}
3	0.188 atm	0.500 atm	1.56 x 10^{-4} atm s^{-1}
4	1.000 atm	1.000 atm	9.00 x 10^{-3} atm s^{-1}

3. What is the rate law equation for this reaction?

   (A) Rate = $k[P_{NO}]$
   (B) Rate = $k[P_{NO}]^2[P_{H_2}]$
   (C) Rate = $k[P_{NO}][P_{H_2}]^2$
   (D) Rate = $k[P_{NO}][P_{H_2}]$
   (E) Rate = $k[P_{NO}]^2$

4. When both the NO and the H₂ each have an initial partial pressure of 2.000 atm, the initial rate of the reaction will be:

(A) 9.00×10^{-3} atm s^{-1}
(B) 1.80×10^{-2} atm s^{-1}
(C) 2.70×10^{-2} atm s^{-1}
(D) 3.60×10^{-2} atm s^{-1}
(E) 7.20×10^{-2} atm s^{-1}

5. The value of the rate constant, k, is:

(A) 9.00×10^{-3} atm s^{-1}
(B) 9.00×10^{-3} atm^{-1} s^{-1}
(C) 9.00×10^{-3} atm^{-2} s^{-1}
(D) 9.00×10^{-4} atm^{3} s
(E) 9.00×10^{-4} atm^{4} s^{-1}

6. Which could be the rate determining step in the mechanism of this reaction?

(A) 2NO + 2H₂ ⇒ Intermediate
(B) 2NO ⇒ Intermediate
(C) NO + 2H₂ ⇒ Intermediate
(D) NO + H₂ ⇒ Intermediate
(E) 2NO + H₂ ⇒ Intermediate

7. Which conclusion can be made from the data given?
The rate determining step:
(A) of a catalyzed reaction is faster.
(B) is not assumed from the stoichiometry of the reaction.
(C) is the fastest step.
(D) is the slowest step.
(E) at a temperature 10°C higher doubles the rate.

Questions 8 and 9 deal with the reaction:
   N₂(g) + O₂(g) ⇒ 2NO(g)
Assume there are *no* intermediate steps or other complications.
Use these choices for *both* questions.

(A) The reaction rate will increase, but not double.
(B) The reaction rate will double.
(C) The reaction rate will more than double.
(D) The reaction rate will decrease, but not halve.
(E) The reaction rate will halve.

8. The temperature is adjusted so the reaction just begins. If the partial pressure of the O₂ is doubled at this temperature:

9. The number of molecules and the volume are held constant. If the temperature is raised high enough to double the number of collisions:

10. Using the data for the reaction: $I^- + OCl^- \Rightarrow IO^- + Cl^-$

Run	$[I^-]$ M	$[OCl^-]$ M	Rate
1	0.07 M	0.21 M	0.0537 mol L^{-1} s^{-1}
2	0.14 M	0.42 M	0.215 mol L^{-1} s^{-1}
3	0.28 M	0.42 M	0.429 mol L^{-1} s^{-1}
4	0.56 M	0.21 M	0.429 mol L^{-1} s^{-1}

Find the rate equation?

(A) Rate = k $[I^-]$
(B) Rate = k $[OCl^-]$
(C) Rate = k $[I^-][OCl^-]$
(D) Rate = k $[I^-]^2[OCl^-]$
(E) Rate = k $[I^-][OCl^-]^2$

11. The energy diagram for a reaction is given in the graph. The mechanism for the reaction is:
Step 1:   A + B   ⇔   C
Step 2:   C + B   ⇒   D
Step 3:   D + A   ⇒   F
Overall: 2A + 2B ⇒ F

Which is true about this system and the data?

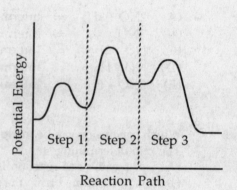

(A) Step 3 is the fastest step.
(B) A catalyst would affect the potential energy of each step equally.
(C) The rate law equation is: Rate = k $[A]^2[B]^2$.
(D) The rate law equation is: Rate = k $[A][B]^2$.
(E) The activation energy of each step is equal.

12. For the reaction: $H_2(g) + Cl_2(g) \Rightarrow 2HCl(g)$
 • Equal volumes of hydrogen and chlorine react explosively.
 • A large excess of either will not explode, but may react.

What will explain this difference?
(A) An increased activation energy.
(B) A decreased heat of reaction.
(C) An increased concentration of the chain–maintaining agent.
(D) An increased concentration of the chain–ending reactant.
(E) A decreased collision probability between chain–maintaining species.

13. The decomposition of $NaClO_3$ is a first order reaction. In an experiment, a sample of $NaClO_3$ is 90% decomposed in 48.0 minutes. How long will it take for a sample to be 50% decomposed?

    (A) 14.4 minutes
    (B) 21.6 minutes
    (C) 24.0 minutes
    (D) 43.2 minutes
    (E) 48.0 minutes

Questions 14 and 15 refer to the graph shown. The solid line (B) is a graph of the kinetic energy distribution at a given temperature. Point E_a represents the activation energy. Use the choices for *both* questions.

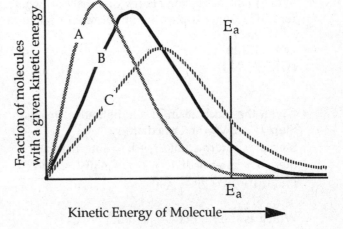

(A) point E_a moves to the right.
(B) point E_a moves to the left.
(C) the curve becomes A.
(D) the curve becomes C.
(E) none of the above.

14. If the temperature is increased:

15. If a catalyst is added:

16. The energy diagram is for the reaction:
    $NO(g) + O_3(g) \Leftrightarrow NO_2(g) + O_2(g)$
    The distance marked 'X' represents:

    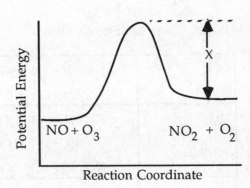

    (A) The activation energy for:
    $NO(g) + O_3(g) \Leftrightarrow NO_2(g) + O_2(g)$
    (B) The heat of reaction for:
    $NO(g) + O_3(g) \Leftrightarrow NO_2(g) + O_2(g)$
    (C) The activation energy for:
    $NO_2(g) + O_2(g) \Leftrightarrow NO(g) + O_3(g)$
    (D) The heat of reaction for:
    $NO_2(g) + O_2(g) \Leftrightarrow NO(g) + O_3(g)$
    (E) None of the above.

Multiple-Choice Questions

17. An approximate rule is that an increase in temperature of 10°C will double the reaction rate. Which is doubled on the molecular level? The:

    (A) activation energy.
    (B) average kinetic energy.
    (C) average velocity of molecules.
    (D) fraction of molecules with the activation energy.
    (E) number of collisions.

18. What is the activation energy of a reaction where an increase in temperature from 27°C to 37°C exactly triples the initial rate of reaction?
    (A) 232 kJ (D) 75.7 kJ
    (B) 172 kJ (E) 43.9 kJ
    (C) 84.9 kJ

19. Given the mechanism of a hypothetical reaction:
    Step 1: $2A \Leftrightarrow intermediate_{(1)}$ (fast equilibrium)
    Step 2: $intermediate_{(1)} + B \Rightarrow intermediate_{(2)}$ (slow)
    Step 3: $intermediate_{(2)} + B \Rightarrow A_2B_2$ (fast)
    Overall: $2A + 2B \Rightarrow A_2B_2$

    Which is the rate law equation?
    (A) rate = k[A]^2 (D) rate = k[A]^2[B]
    (B) rate = k[B]^2 (E) rate = k[A][B]^2
    (C) rate = k[A][B]

20. The reaction $2NO(g) + Cl_2(g) \Rightarrow 2NOCl(g)$ has the rate law:
    Rate = k[NO]^2[Cl$_2$]
    Which is a possible mechanism?

    | (A) | $2NO \Rightarrow N_2O_2$ | slow |
    | | $N_2O_2 + Cl_2 \Rightarrow 2NOCl$ | fast |
    | (B) | $2NO \Leftrightarrow N_2O_2$ | fast equilibrium |
    | | $N_2O_2 \Rightarrow NO_2 + N$ | slow |
    | | $NO_2 + N + Cl_2 \Rightarrow 2NOCl$ | fast |
    | (C) | $Cl_2 \Rightarrow 2Cl$ | slow |
    | | $2NO \Leftrightarrow N_2O_2$ | fast equilibrium |
    | | $N_2O_2 + 2Cl \Rightarrow 2NOCl$ | fast |
    | (D) | $NO + Cl_2 \Leftrightarrow NOCl_2$ | fast equilibrium |
    | | $NO + NOCl_2 \Rightarrow 2NOCl$ | slow |
    | (E) | $NO + Cl_2 \Rightarrow NOCl + Cl$ | slow |
    | | $Cl + NO \Rightarrow NOCl$ | fast |

Answer Key—Set 5

1.	(E)	6.	(E)	11.	(D)	16.	(C)
2.	(A)	7.	(B)	12.	(E)	17.	(D)
3.	(B)	8.	(B)	13.	(A)	18.	(C)
4.	(E)	9.	(C)	14.	(D)	19.	(D)
5.	(C)	10.	(C)	15.	(B)	20.	(D)

Explanations

1. **(E)** There are 2NO molecules and one Br_2, so the rate of disappearance (minus sign is required) of the bromine, Br_2, is $\frac{1}{2}$ the [NO]. The order of [NO] and [Br_2] are 2 and 1 respectively.

2. **(A)** The reaction rate is $-(\frac{1}{2} \times -2.0)$ or 1.0 mmHg s^{-1}. The appearance of O_2 is (3 x 1.0 mmHg s^{-1} = 3.0 mmHg s^{-1}.)

3. **(B)** Runs 1 and 3 show that when P(NO) alone is doubled, the reaction rate is quadrupled. The order of NO is 2. Runs 1 and 2 show that when P(H_2) alone is doubled, the reaction rate is doubled. The order of H_2 is 1.

4. **(E)** The overall order is 3. The rate will be 8 times, $(2)^3$, that of run 4 at the same temperature.
   Rate = 8.000 x 9.00 x 10^{-3} = 7.20 x 10^{-2} atm s-1

5. **(C)** Run 4 shows the rate constant, k, is 9.00 x 10^{-3}.

   $$\text{Rate} = 9.00 \times 10^{-3} \frac{atm}{sec} = k \frac{1}{atm^2 \, sec} (1.0)^2 (1.0) atm^3$$

6. **(E)** The rate determining step is the *slow* step, and the reactants will have orders which are equal to their coefficients in this step.

7. **(B)** The data only shows that the stoichiometry and the orders are different. All the other answers (except (C)) are essentially true, but are not reflected in the data.

8. **(B)** Doubling the pressure will double the collisions and therefore double the reaction rate. **This is an often used rule of thumb.**

9. **(C)** Raising the temperature high enough to double the number of collisions will also increase the effectiveness of the collisions.

Multiple-Choice Questions

10. **(C)** Runs 2 and 3 show that when [I⁻] is doubled, the reaction rate is doubled. The order of I⁻ is 1. Comparing runs 1 and 4 show that [OCl⁻] is 8-times greater and the reaction rate is 8-times greater. The order of OCl⁻ is 1.

11. **(D)** The rate determining step is 2, the step with the *highest* activation energy.
    Rate = $k_2[C][B]$
    Step 1 is a fast equilibrium.
    $[C] = K_1[A][B]$
    Substitute for [C] in the rate equation.
    Rate = $k[A][B]^2$ $k = K_1 k_2$

12. **(E)** Diluting the system puts H_2 or Cl_2 molecules in between intermediate species (free atoms) that are essential for an explosive chain reaction step.

13. **(A)** $\ln 0.10 = -k \times 48.0 \text{ min}$
    $k = 0.0480 \text{ min}^{-1}$
    $t_{1/2} = \ln 0.50 \div 0.0480 \text{ min}^{-1} = $ **14.4 minutes**

14. **(D)** A higher temperature shifts the temperature curve to the right and with a lower height.

15. **(B)** A catalyst always lowers the activation energy, E_a.

16. **(C)** 'X' is the reverse reaction activation energy.

17. **(D)** Increasing the temperature shifts the kinetic energy curve to the right, which increases (and sometimes doubles) the number of molecules with the required activation energy.

18. **(C)** Use the Arrhenius equation.
    $$\ln 3.0 = -\frac{E_a}{8.314 \text{ J mol}^{-1} \text{ K}^{-1}} \left[\frac{1}{310} - \frac{1}{300} \right]$$

19. **(D)** Step 2 is rate determining.
    Rate = $k_2 [\text{Int}_1][B]$
    Step 1 is a fast equilibrium.
    $[\text{Int}_1] = K_1 [A]^2$
    Substitute $K_1 [A]^2$ for $[\text{Int}_1]$ in the rate equation.

20. **(D)** For the slow step:
    Rate = $k_2 [\text{NOCl}_2][\text{NO}]$.
    The fast equilibrium gives:
    $[\text{NOCl}_2] = K_1 [\text{NO}][\text{Cl}_2]$.
    Combining these will agree with the rate equation.

Set 6

Equilibrium

1. For a reaction involving only gases at 25°C., the equilibrium constant can be expressed in terms of molarity, K_c, or partial pressure, K_p. Which is true about the numerical value of K_p?

   (A) K_c is generally greater than K_p.
   (B) K_c is generally less than K_p.
   (C) K_c is generally equal to K_p.
   (D) K_c is equal to K_p if the total moles of reactants and total moles of products are equal.
   (E) K_c is greater than K_p if the total moles of reactants are less than the total moles of products.

2. The equilibrium reaction is:
   $$H_2(g) + I_2(g) \Leftrightarrow 2HI(g)$$
   When 2.00 mole each of hydrogen, H_2, and iodine, I_2, are mixed in a 1.00 L vessel, 3.50 mole of HI is produced.
   What is the value of the equilibrium constant, K_c?

   (A) 196
   (B) 56
   (C) 49
   (D) 14
   (E) 5.4

3. The equilibrium constant for the reaction:
   $$SO_2(g) + NO_2(g) \Leftrightarrow SO_3(g) + NO(g)$$
   has a numerical value of 3.00 at a given temperature. Equimolar amounts SO_2 and NO_2 are mixed at this temperature and a total pressure of 3.00 atm.
   What percent of the SO_2 is converted to product?

   (A) 87.5%
   (B) 74.0%
   (C) 63.4%
   (D) 50.0%
   (E) 33.0%

4. For which of the reactions will an increase in pressure cause a decrease in product (temperature remaining constant)?

   (A) $N_2(g) + 3H_2(g) \Leftrightarrow 2NH_3(g)$
   (B) $3Fe(s) + 4H_2O(g) \Leftrightarrow Fe_3O_4(s) + 4H_2(g)$
   (C) $PCl_3(g) + Cl_2(g) \Leftrightarrow PCl_5(g)$
   (D) $HCl(g) + H_2O(l) \Leftrightarrow H_3O^+(aq) + Cl^-(aq)$
   (E) $CaCO_3(s) \Leftrightarrow CaO(s) + CO_2(g)$

5. Consider the equilibrium reaction:
   $2SO_2(g) + O_2(g) \Leftrightarrow 2SO_3(g)$
   What will be the effect of doubling the concentration of SO_3?
   The concentration of:

   (A) SO_2 and O_2 increase equally.
   (B) SO_2 increases more than that of O_2.
   (C) O_2 increases more than that of SO_2.
   (D) SO_2 decreases more than that of O_2.
   (E) O_2 decreases more than that of SO_2.

6. Barium hydroxide is completely ionized in dilute solution.
   $Ba(OH)_2(s) \Rightarrow Ba^{2+}(aq) + 2OH^-(aq)$
   Which is true?

   (A) $[Ba^{2+}] = [OH^-]^2$
   (B) $[Ba^{2+}] = 2 \times [OH^-]$
   (C) $[OH^-] = [Ba^{2+}]^2$
   (D) $[OH^-] = 2 \times [Ba^{2+}]$
   (E) $[Ba^{2+}] = [Ba(OH)_2]$

7. The solubility of copper II iodate, $Cu(IO_3)_2$ in water is 3.27×10^{-3} M. What is the K_{sp}?

   (A) 2.14×10^{-5}
   (B) 4.28×10^{-5}
   (C) 1.40×10^{-7}
   (D) 3.50×10^{-8}
   (E) 7.00×10^{-8}

8. A 10.0 mL solution is 0.10 M with respect to both calcium nitrate, $Ca(NO_3)_2$, and magnesium nitrate, $Mg(NO_3)_2$. What happens if 0.10 M NaF is added by drops until precipitation occurs?
   $K_{sp} = 4.0 \times 10^{-11}$ for CaF_2 at 25°C.
   $K_{sp} = 6.4 \times 10^{-9}$ for MgF_2 at 25°C.

   (A) The precipitate is calcium fluoride only.
   (B) The precipitate is magnesium fluoride only.
   (C) The precipitate contains both CaF_2 and MgF_2.
   (D) The precipitate contains both $Ca(NO_3)_2$ and MgF_2.
   (E) The precipitate contains both CaF_2 and $Mg(NO_3)_2$.

9. Why is hydroxide ion a strong base in water solutions?

   (A) OH^- is the only base.
   (B) OH^- is the conjugate base of the hydronium ion, H_3O^+.
   (C) All bases react with water to produce OH^- ions.
   (D) All acids react completely with OH^- ions.
   (E) Strong bases dissolve completely in water and produce OH^- ions.

Equilibrium

10. Which is true about a 0.25 M KOH solution?

    (A) [KOH] = 0.25 M
    (B) [K$^+$] = [OH$^-$] = 0.25 M
    (C) [KOH] = [K$^+$] = 0.25 M
    (D) [H$^+$] = [OH$^-$] = 1.0 × 10^{-7} M
    (E) pH = 0.60

11. The ionization reaction for water is:
    $$2H_2O(l) \Leftrightarrow H_3O^+(aq) + OH^-(aq)$$
    Which is a true statement about this reaction?

    (A) If water is added the [H$_3$O$^+$] and [OH$^-$] increase.
    (B) If acid is added the [H$_3$O$^+$] and [OH$^-$] decrease.
    (C) If base is added the [H$_3$O$^+$] and [OH$^-$] increase.
    (D) $K_w = \dfrac{[H_3O^+][OH^-]}{[H_2O]^2}$
    (E) In pure water [H$_3$O$^+$] equals [OH$^-$].

12. To what total volume of aqueous solution must 20.0 mL of 0.10 M HCl be diluted so that the pH is 1.30?

    (A) 20. mL
    (B) 25. mL
    (C) 40. mL
    (D) 50. mL
    (E) 110 mL

13. The term "K_a for the ammonium ion, NH$_4^+$," refers to which equation?

    (A) NH$_4^+$(aq) + H$_2$O(l) $\Leftrightarrow$ NH$_3$(aq) + H$_3$O$^+$(aq)
    (B) NH$_4^+$(aq) + OH$^-$(aq) $\Leftrightarrow$ NH$_3$(aq) + H$_2$O(l)
    (C) NH$_3$(aq) + H$_2$O(l) $\Leftrightarrow$ NH$_4^+$(aq) + OH$^-$(aq)
    (D) NH$_3$(aq) + H$_3$O$^+$(aq) $\Leftrightarrow$ NH$_4^+$(aq) + OH$^-$(aq)
    (E) None of these.

14. The K_a of hydrocyanic acid, HCN, is 5.0 × 10^{-10}. The pH of 0.05 M HCN(aq) is:

    (A) between 3.5 and 4.5
    (B) between 4.5 and 5.0
    (C) between 5.0 and 5.5
    (D) between 5.5 and 6.0
    (E) between 9.0 and 9.5

15. Benzoic acid, C$_6$H$_5$COOH, is 1.0% ionized in a 0.010 M solution.
    $$C_6H_5COOH(aq) \Leftrightarrow H^+(aq) + C_6H_5COO^-(aq)$$

    What is the value of K_a?
    (A) 1.0 × 10^{-8}
    (B) 1.0 × 10^{-6}
    (C) 1.0 × 10^{-4}
    (D) 1.0 × 10^{-3}
    (E) 1.0 × 10^{-2}

16. Which set of solutes will form a buffer when dissolved in water to make 1 liter of solution?

(A) 0.2 mole of NaOH with 0.2 mole of HCl.
(B) 0.2 mole of NaCl with 0.2 mole of HNO_3.
(C) 0.4 mole of CH_3COOH with 0.4 mole of NaOH.
(D) 0.4 mole of NH_3 with 0.2 mole of HCl.
(E) 0.3 mole of KOH with 0.2 mole of HBr.

17. An experiment is performed where it is desired to keep the pH at 3.70. Which system would you use?

(A) 0.0002 M HCl
(B) HCNO and CNO^- with a $K_a = 2.19 \times 10^{-4}$
(C) CH_3COOH and CH_3COO^- with a $K_a = 1.74 \times 10^{-5}$
(D) HOCl and OCl^- with a $K_a = 2.95 \times 10^{-8}$
(E) C_6H_5OH and $C_6H_5O^-$ with a $K_a = 5.00 \times 10^{-11}$

18. The K_a for chloroacetic acid, CH_2ClOOH, is 1.35×10^{-3}. What is the pH of a solution that is 0.15 M in CH_2ClOOH and 0.25 M in sodium chloroacetate, $CH_2ClOONa$?

(A) 1.17
(B) 2.64
(C) 2.87
(D) 3.09
(E) 4.54

19. What is the pH of a solution obtained by mixing 10. mL of 1.0 M HCl, 75 mL of water and 15 mL of 1.0 M NaOH?

(A) 1.3
(B) 1.7
(C) 7.0
(D) 12.3
(E) 12.7

20. What is the pH of a solution obtained by titrating 50.0 mL of 0.100 M HNO_2 with 0.100 M NaOH to the equivalence point? The K_a of HNO_2 is 4.5×10^{-4}.

(A) 1.67
(B) 2.35
(C) 3.35
(D) 7.00
(E) 8.02

Answer Key—Set 6

1.	(D)	6.	(D)	11.	(E)	16.	(D)
2.	(A)	7.	(C)	12.	(C)	17.	(B)
3.	(C)	8.	(A)	13.	(A)	18.	(D)
4.	(E)	9.	(E)	14.	(C)	19.	(E)
5.	(B)	10.	(B)	15.	(B)	20.	(E)

Explanations

1. **(D)** $K_c = K_p(1/RT)^{\Delta n}$ or $K_p = K_c(RT)^{\Delta n}$
   When $\Delta n = 0$ the $(1/RT)$ or the (RT) term becomes equal to one (1), and $K_c = K_p$

2. **(A)**

	$H_2(g)$	+ $I_2(g)$	⇔ 2HI(g)
Start	2.00 M	2.00 M	0.0 M
Δ	−1.75	−1.75	+3.50
Finish	0.25	0.25	3.50

   $$K_c = \frac{(3.50)^2}{(0.25)^2} = 196$$

3. **(C)**

	$CO_2(g)$	+ $H_2(g)$	⇔ CO(g)	+ $H_2O(g)$
Start	1.50 atm	1.50 atm	0.0 atm	0.0 atm
Δ	−x	−x	+x	+x
Finish	1.50−x	1.50−x	x	x

   $$\frac{x^2}{(1.50-x)^2} = 3.00$$
   $x = 0.951$
   $$\% = \frac{0.951}{1.50} \times 100 = \mathbf{63.4\%}$$

4. **(E)** When pressure is increased the side of the reaction with the least moles of gas is favored. Reaction E has 0 moles of gaseous reactant and 1 mole of products. (E) will produce reactant at the expense of product as it returns to equilibrium.

5. **(B)** Increasing the concentration of a product shifts the equilibrium to the left. The concentrations of SO_2 and O_2 increase by stoichiometric amounts.

6. **(D)** The stoichiometry of the balanced equation predicts that for every barium ion (Ba^{2+}) formed two (2) hydroxide ions, OH^- are produced.

7. **(C)** $Cu(IO_3)_2(s) \Leftrightarrow Cu^{2+}(aq) + 2IO_3^-(aq)$

   $K_{sp} = [Cu^{2+}][IO_3^-]^2 = [3.27 \times 10^{-3}][6.54 \times 10^{-3}]^2$
   $K_{sp} = 1.40 \times 10^{-7}$

8. **(A)** Only the least soluble fluoride precipitates first.
   $MF_2(s) \Leftrightarrow M^{2+}(aq) + 2F^-(aq)$
   where M^{2+} is Ca^{2+} or Mg^{2+}

   $K_{sp} = [M^{2+}][F^-]^2 = [0.10][F^-]^2$
   $[F^-] = (K_{sp} \div 0.10)^{1/2}$
   The CaF_2 preciptates when $[F^-] = 2.0 \times 10^{-5}$ M.
   The MgF_2 preciptates when $[F^-] = 2.5 \times 10^{-4}$ M.

9. **(E)** Main Group 1 hydroxides are strong bases which dissociate completely in aqueous solutions to produce alkali metal ions and hydroxide ions. Other bases react with water to form hydroxide ions.

10. **(B)**

	KOH(s)	$\Leftrightarrow K^+(aq)$	$+ OH^-(aq)$
Start	0.25 M	0.0 M	0.0 M
Δ	−0.25	+0.25	+0.25
Finish	0.0	0.25	0.25

    The $[K^+(aq)]$ and $[OH^-(aq)]$ are equal at 0.25 M.

11. **(E)** $2H_2O(l) \Leftrightarrow H_3O^+(aq) + OH^-(aq)$
    At equilibrium: $[H_3O^+(aq)] = [OH^-(aq)] = 1.0 \times 10^{-7}$ M
    or in the simplified format: $[H^+] = [OH^-] = 1.0 \times 10^{-7}$ M

12. **(C)** When the pH is 1.30, the $[H^+] = 10^{-1.30} = 0.050$ M.
    This is a dilution problem and the equation for dilution should be used.
    $M_2V_2 = M_1V_1$
    $0.050 \text{ M} \times V_2 = 0.10 \text{ M} \times 20.0 \text{ mL}$
    $V_2 = 40. \text{ mL}$

Equilibrium 139

13. **(A)** K_a is called the dissociation constant and refers to the reaction where ions are formed when an acid dissolved in water.
    Another way to write the K_a equation is:
    $$NH_4^+(aq) \Leftrightarrow H^+(aq) + NH_3(aq)$$
    H_3O^+, the hydronium ion, is the same as a hydrated proton, $H^+(aq)$.
    $$H_3O^+ = H^+(aq)$$

14. **(C)**

	HCN(aq)	$\Leftrightarrow$ H$^+$(aq)	+ CN$^-$(aq)
Finish	0.05 M	x M	x M

    $$K_a = \frac{x^2}{0.05} = 5 \times 10^{-6}$$
    $$pH = -\log[H^+] = 5.3$$

15. **(B)**

	C$_6$H$_5$COOH	$\Leftrightarrow$ H$^+$	+ C$_6$H$_5$COO$^-$
Finish	0.0100 M	1.00×10^{-4} M	1.00×10^{-4} M

    $$K_a = \frac{(1.00 \times 10^{-4})^2}{1.00 \times 10^{-2}} = 1.0 \times 10^{-6}$$

16. **(D)** $NH_4^+(aq) \Leftrightarrow H^+(aq) + NH_3(aq)$
    When acid is added to NH_3, some $NH_4^+(aq)$ is formed. A buffer is roughly equivalent concentrations of a weak acid and its conjugate base.

17. **(B)** The pH is 3.70 and the $[H^+] = 10^{-3.70} = 2.0 \times 10^{-4}$.
    $$K_a = \frac{[H^+] \times [HA]}{[A^-]} = [H^+] = 2.0 \times 10^{-4}.$$
    The HCNO and CNO$^-$ buffer system has this K_a when [HCNO] = [CNO$^-$].

18. **(D)** The Henderson-Hasselbalch equation can be used to calculate the pH of a buffer.
    $$pH = pK_a + \log\frac{[A^-]}{[HA]} = 2.87 + \log\frac{(0.25)}{(0.15)} = 3.09$$

19. **(E)** $[OH^-] = \dfrac{15\ mL \times 1.0\ mol\ L^{-1}}{100.\ mL} = 0.015\ M$

   $[H^-] = \dfrac{10\ mL \times 1.0\ mol\ L^{-1}}{100.\ mL} = 0.010\ M$

	$H_2O(l)$	$\Leftrightarrow H^+(aq)$	$+ OH^-(aq)$
Start	Some	0.10 M	0.15 M
Δ	+0.10	−0.10	−0.10
Finish	Some	0.00	0.05

$[H^+] = \dfrac{K_w}{[OH^-]} = \dfrac{1.0 \times 10^{-14}}{0.05} = 2 \times 10^{-13}$

$pH = -\log[H^+] = \mathbf{12.7}$

20. **(E)** Both reagents are 0.100 M. At the equivalence point, 50.0 mL of base has been added to 50.0 mL of acid.

   $[HNO_2] = [OH^-] = \dfrac{50.0\ mL \times 0.100\ mol\ L^{-1}}{100.\ mL} = 0.0500\ M$

   The titration reaction is:
   $$HNO_2 + OH^-(aq) \Rightarrow NO^-(aq) + H_2O(l)$$
   The reaction has produced 0.0500 M $NO^-(aq)$, a base.
   pH = 8.02 is the only base pH in the answer set.

   The base reaction is the reverse of the neutralization reaction.

	$NO^-(aq)$	$+ H_2O(l)$	$\Rightarrow HNO_2(aq)$	$+ OH^-(aq)$
Finish	0.0500	Some	x	x

$K_b = \dfrac{K_w}{K_a} = \dfrac{1.0 \times 10^{-14}}{4.5 \times 10^{-4}} = 2.2 \times 10^{-11}$

$K_b = \dfrac{[HNO_2][OH^-]}{[NO^-]} = \dfrac{x^2}{0.0500} = 2.2 \times 10^{-11}$

$x = [OH^-] = 1.05 \times 10^{-6}\ M$

$[H^+] = \dfrac{K_w}{[OH^-]} = \dfrac{1.00 \times 10^{-14}}{1.05 \times 10^{-6}} = 9.52 \times 10^{-9}$

$pH = -\log[H^+] = \mathbf{8.02}$

Set 7
Thermodynamics

1. A gas evolves 1420. J. of heat as it is compressed from 30.0 L to 10.0 L by an opposing pressure of 2.00 atm.
   Note: 1.00 Liter-atm = 101 J.
   The internal energy change, ΔE, in *joules* for the gas is:

   (A) +5460 J
   (B) −5460 J
   (C) +1420 J
   (D) −2620 J
   (E) +2620 J

2. In which of the reactions would enthalpy change, ΔH, be most nearly equal to the internal energy change, ΔE?

   (A) $2H_2(g) + O_2(g) \Rightarrow 2H_2O(l)$
   (B) $H_2(g) + Cl_2(g) \Rightarrow 2\ HCl(g)$
   (C) $C_2H_5Cl(g) \Rightarrow C_2H_4(g) + HCl(g)$
   (D) $2CO(g) + O_2(g) \Rightarrow 2CO_2(g)$
   (E) $C_2H_4(g) + 3O_2(g) \Rightarrow 2CO_2(g) + 2H_2O(l)$

3. If the heat of reaction, $\Delta H°$, is −350 kJ at 25°C for the process:
   $HCl(aq) + CaCO_3(s) \Rightarrow CaCl_2(aq) + H_2O(l) + CO_2(g)$
   Which of the following numerical values must be substituted into the equation to find the internal energy change, $\Delta E°$, in kJ?
   $\Delta H° = \Delta E° + \Delta nRT$

   (A) −298
   (B) 1.987×10^{-3}
   (C) 0.0821
   (D) 1
   (E) 623

4. The combustion of 0.100 g of ethane (C_2H_6) caused the temperature of a bomb calorimeter and its contents to rise 1.74°C. The calorimeter and the water it contains have a heat capacity of 2980 J °C^{-1}. What is the energy change, in kJ mol^{-1}, for the combustion of ethane?

   (A) ΔE = −5190 kJ mol^{-1}
   (B) ΔH = −5190 kJ mol^{-1}
   (C) ΔE = −1560 kJ mol^{-1}
   (D) ΔH = −1560 kJ mol^{-1}
   (E) ΔE = − 515 kJ mol^{-1}

5. When 2.00 grams of hydrogen fluoride gas, HF (MM = 20.0), is bubbled into 400. mL of water at 20.6°C, all the gas dissolves, and the temperature of the solution rises to 24.3°C. The density and specific heat (4.184 J g^{-1} °C^{-1}) of the solution may be assumed to be the same as that of pure water. What is the value of the enthalpy change of solution, $\Delta H_{solution}$, for HF?

(A) − 62   kJ mol^{-1}
(B) − 41   kJ mol^{-1}
(C) − 6.2 kJ mol^{-1}
(D) + 6.2 kJ mol^{-1}
(E) + 62   kJ mol^{-1}

6. Find $\Delta H°$ for the reaction, in kJ.
$$3V_2O_3(s) \Rightarrow V_2O_5 + 4VO(s)$$
Standard heat of reactions, $\Delta H°$, are given in the table.

Reaction			$\Delta H°$, kJ
4VO(s)	+ O$_2$(g)	$\Rightarrow$ 2V$_2$O$_3$(s)	−753
2V$_2$O$_3$(s)	+ O$_2$(g)	$\Rightarrow$ 4VO$_2$(s)	−452
4VO$_2$(s)	+ O$_2$(g)	$\Rightarrow$ 2V$_2$O$_5$(s)	−243

(A) + 858 kJ
(B) + 405 kJ
(C) + 59 kJ
(D) − 649 kJ
(E) −1448 kJ

7. Given the thermochemical reaction data:

Reaction				$\Delta H°$, kJ
MnO$_2$(s)	+ CO(g)	$\Rightarrow$ MnO(s)	+ CO$_2$(g)	−151
Mn$_3$O$_4$(s)	+ CO(g)	$\Rightarrow$ 3MnO(s)	+ CO$_2$(g)	− 54
3Mn$_2$O$_3$(s)	+ CO(g)	$\Rightarrow$ 2Mn$_3$O$_4$(s)	+ CO$_2$(g)	−142

Find $\Delta H°$, in kJ, for the reaction:

$$2MnO_2(s) + CO(g) \Rightarrow Mn_2O_3(s) + CO_2(g)$$

(A) − 385 kJ
(B) − 347 kJ
(C) − 313 kJ
(D) − 219 kJ
(E) − 45 kJ

8. What is the heat of reaction, $\Delta H°$, in kJ mol^{-1} for the reaction?
$$CH_4(g) + 4Cl_2(g) \Rightarrow CCl_4(l) + 4HCl(g)$$

Heat of formation, $\Delta H_f°$, data is given in the table.

Substance	$\Delta H_f°$, kJ mol^{-1}
$CH_4(g)$	−74.9
$CCl_4(l)$	−135.6
$HCl(g)$	−92.5

(A) The heat of reaction, $\Delta H°$, cannot be computed without additional data.
(B) − 153 kJ
(C) − 303 kJ
(D) − 430 kJ
(E) − 580 kJ

9. The standard heat of combustion, $\Delta H_c°$, for the reaction is −1985 kJ.
$$2(CH_3)_2CHOH(l) + 9O_2(g) \Rightarrow 6CO_2(g) + 8H_2O(l)$$

Heat of formation, $\Delta H_f°$, data is given in the table.

Substance	$\Delta H_f°$, kJ mol^{-1}
$CO_2(g)$	−394
$H_2O(l)$	−286
$(CH_3)_2CHOH(l)$	?

What is the standard heat of formation, $\Delta H_f°$, of isopropyl alcohol, $(CH_3)_2CHOH(l)$, in kJ mol^{-1}?

(A) +1310 kJ mol^{-1}
(B) − 678 kJ mol^{-1}
(C) −1330 kJ mol^{-1}
(D) −1990 kJ mol^{-1}
(E) −4650 kJ mol^{-1}

10. The standard heat of formation, $\Delta H_f°$, of methane, CH_4, is −75 kJ mol^{-1}.
$$C_{graphite} + 2H_2(g) \Rightarrow CH_4(g)$$

Bond	Bond Energy, kJ mol^{-1}
C − C bonds in graphite	715
H−H bonds in hydrogen	435

Calculate the bond energy of a C −H bond.
(A) 268 kJ mol^{-1}
(B) 305 kJ mol^{-1}
(C) 326 kJ mol^{-1}
(D) 377 kJ mol^{-1}
(E) 415 kJ mol^{-1}

11. The entropy of a perfect crystal is zero when the :

    (A) system is at equilibrium.
    (B) enthalpy change is zero.
    (C) entropy change is zero.
    (D) free energy change is zero.
    (E) system is at absolute zero.

12. Which reaction has the largest increase in entropy, ΔS?

    (A) $CO_2(s) \Leftrightarrow CO_2(g)$
    (B) $H_2(g) + Cl_2(g) \Leftrightarrow 2HCl(g)$
    (C) $C_{diamond} \Leftrightarrow C_{graphite}$
    (D) $2H_2(g) + O_2(g) \Leftrightarrow H_2O(g)$
    (E) $KNO_3(s) \Leftrightarrow KNO_3(l)$

13. The heat of fusion of sodium is 2610 J mol^{-1} at its normal melting temperature of 98.0 °C.
    $$Na(s) \Leftrightarrow Na(l)$$
    What is the entropy change, $\Delta S°$, in J mol^{-1} K^{-1} at this temperature?

    (A) +1550
    (B) +26.5
    (C) +7.04
    (D) +2.50
    (E) +0.661

14. Which is true for a spontaneous endothermic process?

    (A) $\Delta H < 0$
    (B) $S° = 0$
    (C) $\Delta S > 0$
    (D) $\Delta G > 0$
    (E) $\Delta G = 0$

15. Which reaction will be spontaneous?

	ΔH, kJ	ΔS, J K^{-1}	T, K
(A)	+50	-50	1000
(B)	+250	+400	300
(C)	+250	+50	300
(D)	-100	-400	250
(E)	-100	-400	200

16. A reaction has values of ΔH and ΔS which are both positive. The reaction:
    (A) cannot be spontaneous.
    (B) spontaneity is temperature dependent.
    (C) is exothermic.
    (D) has an increasing free energy.
    (E) must be at equilibrium.

Questions 17 and 18 deal with the reaction:

$$N_2O_5(g) \Leftrightarrow 2NO_2(g) + 1/2 O_2(g)$$

Substance	$\Delta H_f°$ kJ mol^{-1}	$\Delta G_f°$ kJ mol^{-1}	S° J mol^{-1}
$N_2O_5(g)$	11.3	115.1	355.6
$NO_2(g)$	33.2	51.30	240.0
$O_2(g)$	------	------	205.0

17. What is the $\Delta G°$ of the reaction?
    (A) + 166.4 kJ
    (B) + 115.1 kJ
    (C) + 63.8 kJ
    (D) − 12.5 kJ
    (E) − 66.1 kJ

18. Use the equation:
    $$\Delta G° = \Delta H° - T\Delta S°$$
    to calculate the value of ΔG, in kJ mol^{-1}, at 373 K.
    (A) − 29.5 kJ mol^{-1}
    (B) − 84.5 kJ mol^{-1}
    (C) − 12.6 kJ mol-1
    (D) − 171.2 kJ mol^{-1}
    (E) + 54.8 kJ mol^{-1}

19. A gaseous reaction has a $\Delta G°$ of −24.1 kJ mol^{-1}.
    What is the approximate value of K_p at 298 K?
    (A) 3×10^8
    (B) 2×10^4
    (C) 1.0
    (D) 3×10^{-3}
    (E) 6×10^{-5}

20. The standard net cell potential, $\mathcal{E}°$, is +1.50 volts for the reaction:
    $$Au^{3+}(aq) + 3e^- \Leftrightarrow Au(s)$$
    What is the standard free energy, $\Delta G°$, in kJ?
    (A) − 606
    (B) − 434
    (C) − 145
    (D) − 34.5
    (E) 0

Answer Key—Set 7

1.	(E)	6.	(B)	11.	(E)	16.	(B)
2.	(B)	7.	(D)	12.	(A)	17.	(D)
3.	(D)	8.	(D)	13.	(C)	18.	(A)
4.	(C)	9.	(C)	14.	(C)	19.	(B)
5.	(A)	10.	(E)	15.	(E)	20.	(D)

Explanations

1. **(E)** $\Delta E = q + w$
   $\Delta E = -1420 \text{ J} + (2 \text{ atm}) \times (20 \text{ L}) \times (101 \text{ J L}^{-1} \text{ atm}^{-1})$
   $\Delta E = \mathbf{+2620 \text{ J}}$

2. **(B)** $\Delta H = \Delta E + \Delta nRT$
   $\Delta n = \text{mol of (g)}_{products} - \text{mol of (g)}_{reactants} = 0$ (for reaction (B))

3. **(D)** $\Delta H = \Delta E + \Delta nRT$
   $$-350. \text{ kJ} = \Delta E + \frac{(1 \text{ mol}) \times (8.314 \text{ J mol}^{-1} \text{ K}^{-1}) \times (298 \text{ K})}{1000 \text{ J kJ}^{-1}}$$

4. **(C)** $q_v = (2980 \text{ J °C}^{-1}) \times (1.74 \text{ °C}) = -5190 \text{ J} = -5.19 \text{ kJ}$
   $n = 0.100 \text{ g} \div 30.0 \text{ g mol}^{-1} = 0.00333 \text{ mol}$
   $\Delta E = q_v$
   $\Delta E = -5.19 \text{ kJ} \div 0.00333 \text{ mol} = \mathbf{-1560 \text{ kJ mol}^{-1}}$

5. **(A)** $q_p = (4.184 \text{ J g}^{-1} \text{ °C}^{-1}) \times (400. \text{ g}) \times (24.3 - 20.6)$
   $q_p = -6200 \text{ J} = -6.2 \text{ kJ}$ (minus since heat is *evolved*)
   $n_{HF} = 2.00 \text{ g} \div 20.0 \text{ g mol}^{-1} = 0.100 \text{ mol}$
   $\Delta H = 6.190 \text{ kJ} \div 0.100 \text{ mol} = \mathbf{62 \text{ kJ mol}^{-1}}$

6. **(B)** Use Hess' Law

Reaction			$\Delta H°$, kJ
$2V_2O_3(s)$	$\Rightarrow 4VO(s)$	$+ O_2(g)$	$+753$
$V_2O_3(s)$	$+ \frac{1}{2} O_2(g)$	$\Rightarrow 2VO_2(s)$	$\frac{1}{2} \times -452 = -226$
$2VO_2(s)$	$+ \frac{1}{2} O_2(g)$	$\Rightarrow V_2O_5(s)$	$\frac{1}{2} \times -243 = -122$

$3V_2O_3(s) \Rightarrow V_2O_5 + 4VO(s)$ $\Delta H° = \Sigma \Delta H° = \mathbf{+405 \text{ kJ}}$

Thermodynamics   147

7. **(D)** Use Hess' Law

Reaction	$\Delta H°$, kJ
$2MnO_2(s) + 2CO(g) \Rightarrow 2MnO(s) + 2CO_2(g)$	$2 \times -151 = -302$
$2MnO(s) + \frac{2}{3} CO_2(g) \Rightarrow \frac{2}{3} Mn_3O_4(s) + \frac{2}{3} CO(g)$	$\frac{2}{3} \times +54 = +36$
$\frac{2}{3} Mn_3O_4(s) + \frac{1}{3} CO_2(g) \Rightarrow Mn_2O_3(s) + \frac{1}{3} CO(g)$	$\frac{1}{3} \times +142 = +47$

$2MnO_2(s) + CO(g) \Rightarrow Mn_2O_3(s) + CO_2(g) \quad \Delta H° = \Sigma \Delta H° = -219$ kJ

8. **(D)** $\Delta H° = \Sigma \Delta H_f°$ (Products) $- \Sigma \Delta H_f°$ (Reactants)
    $\Delta H° = ((-135.6) + (4 \times -92.5)) - (-74.9)$
    $\Delta H° = $ **–430. kJ**

9. **(C)** $\Delta H° = \Sigma \Delta H_f°$ (Products) $- \Sigma \Delta H_f°$ (Reactants)
    $\Delta H° = -1985 = ((6 \times -394) + (8 \times -286)) - (2x)$
    $x = \Delta H_f°(CH_3)_2CHOH = $ **–1330 kJ**

10. **(E)** $\Delta H° = $ (Bonds broken) + (Bonds formed)
    Bonds broken $= (+715) + (2 \times (+434))$
    Bonds broken $= +1585$ kJ
    Bonds formed $= (4 \times (- kJ_{C-H}))$
    $\Delta H° = -75$ kJ $= +1585 + (4 \times (- kJ_{C-H}))$
    $kJ_{C-H} = $ **415 kJ mol^{-1}**

11. **(E)** The third law of thermodynamics states that the entropy of a perfect crystal at absolute zero is 0.

12. **(A)** The change from (s) $\Rightarrow$ (g) will result in the highest increase in randomness or chaos.

13. **(C)** $\Delta G = \Delta H - T\Delta S$
    At equilibrium, $\Delta G° = 0$ and $\Delta H° = T\Delta S°$
    $(2610$ J$) = (371$ K$) \times (\Delta S)$
    $\Delta S = $ **7.04 J K^{-1}**

14. **(C)** When a reaction is endothermic, $\Delta H > 0$. For the reaction to be spontaneous, ΔG must be negative ($\Delta G < 0$) and ΔS must therefore be positive ($\Delta S > 0$)

15. **(E)** $\Delta G = \Delta H - \dfrac{T\Delta S}{1000}$

    $\Delta G = (-100 \text{ kJ}) - \dfrac{(200 \text{ K}) \times (-400 \text{ J K}^{-1})}{1000 \text{ J kJ}^{-1}} = -20 \text{ kJ}$

    When $\Delta G < 0$, the reaction is spontaneous.

16. **(B)** At high temperatures, a positive (+) and favorable $T\Delta S$ may exceed a positive (+) and unfavorable ΔH to make ΔG less than zero and the reaction spontaneous.

17. **(D)**
    $\Delta G° = \sum \Delta G_f° \text{ (Products)} - \sum \Delta G_f° \text{ (Reactants)}$
    $\Delta G° = ((2 \times 51.30) + (0)) - (115.1))$
    $\Delta G° = -12.5 \text{ kJ}$

18. **(A)**
    $\Delta H° = \sum \Delta H_f \text{ (Products)} - \sum \Delta H_f \text{ (Reactants)}$
    $\Delta H° = ((2 \times 33.2) + (0)) - (11.3)$
    $\Delta H° = +55.1 \text{ kJ}$

    $\Delta S° = \sum S_f \text{ (Products)} - \sum S_f \text{ (Reactants)}$
    $\Delta S° = ((2 \times 240.0) + (½ \times 205.0)) - (355.6)$
    $\Delta S° = +226.9 \text{ J K}^{-1}$

    $\Delta G° = \Delta H - \dfrac{T\Delta S}{1000}$

    $\Delta G° = (+55.1 \text{ kJ}) - \dfrac{(373 \text{ K}) \times (226.9 \text{ J K}^{-1})}{1000 \text{ J kJ}^{-1}}$

    $\Delta G° = -29.5 \text{ kJ}$

19. **(B)**
    $\Delta G° = -RT \ln K$
    $(-24{,}100 \text{ J mol}^{-1}) = -(8.31 \text{ J mol}^{-1} \text{ K}^{-1}) \times (298 \text{ K}) \times \ln K$
    $\ln K = 9.72$
    $K = e^{9.72}$ (inverse ln K) $= 2 \times 10^4$

20. **(D)**
    $\Delta G° = -nF\mathcal{E}°$
    $\Delta G° = -(3 \text{ mol}) \times (96{,}500 \text{ J mol}^{-1} \text{ volt}^{-1}) \times (1.50 \text{ volt})$
    $\Delta G° = -434{,}000 \text{ J} = -434 \text{ kJ}$

Set 8

Electrochemistry

Note:
Some of the problems require that you consult a table of 'Standard Reduction Potentials'. A table is found in Chapter 8 of Part I, "Review of Selected Topics," and another is provided in the Appendix.

1. Which equation correctly represents balanced reaction for the change from ClO_3^- to $HClO_2$ in acidic solution?

   (A) $ClO_3^- + 3H^+ \Rightarrow HClO_2 + H_2O + 2e^-$
   (B) $ClO_3^- + 2H_2O + 2e^- \Rightarrow HClO_2 + 3OH^-$
   (C) $ClO_3^- + 2H_2O + 2e^- \Rightarrow HClO_2 + 2OH^-$
   (D) $ClO_3^- + 2H_2O \Rightarrow HClO_2 + 3OH^- + 2e^-$
   (E) $ClO_3^- + 3H^+ + 2e^- \Rightarrow HClO_2 + H_2O$

2. Which balanced equation represents the change from $S_2O_3^{2-}$ to SO_3^{2-} in basic solution?

   (A) $S_2O_3^{2-} + 6OH^- \Rightarrow 2SO_3^{2-} + 3H_2O + 4e^-$
   (B) $S_2O_3^{2-} + 6OH^- + 4e^- \Rightarrow 2SO_3^{2-} + 3H_2O$
   (C) $S_2O_3^{2-} + 2e^- \Rightarrow 2SO_3^{2-}$
   (D) $S_2O_3^{2-} + 3H_2O + 4e^- \Rightarrow 2SO_3^{2-} + 6H^+$
   (E) $S_2O_3^{2-} + 3H_2O \Rightarrow 2SO_3^{2-} + 6H^+ + 4e^-$

3. Which is the correct oxidation half-reaction for the reaction between Na_2O_2 and $HSnO_2^-$ to give $HSnO_3^-$ and OH^- in basic solution?

   (A) $Na_2O_2 + 2H_2O + 2e^- \Rightarrow 2Na^+ + 4OH^-$
   (B) $Na_2O_2 + 2H_2O \Rightarrow 2Na^+ + 4OH^- + 2e^-$
   (C) $Na_2O_2 + HSnO_2^- + H_2O \Rightarrow HSnO_3^- + 2Na^+ + 2OH^-$
   (D) $HSnO_2^- + 2OH^- + 2e^- \Rightarrow HSnO_3^- + H_2O$
   (E) $HSnO_2^- + 2OH^- \Rightarrow HSnO_3^- + H_2O + 2e^-$

4. This equation is correctly balanced in basic solution. Which is the oxidizing agent for the reaction as written?
   $8Al + 5OH^- + 3NO_3^- + 18H_2O \Rightarrow 8Al(OH)_4^- + 3NH_3$

   (A) Al
   (B) OH^-
   (C) NO_3^-
   (D) $Al(OH)_4^-$
   (E) NH_3

Multiple-Choice Questions

5. Which can be reduced by good oxidizing agents and oxidized by good reducing agents?

    (A) NH_3
    (B) NO_3^-
    (C) I^-
    (D) SO_3^{2-}
    (E) MnO_4^-

6. An 1.0 M aqueous solution of sodium nitrate, $NaNO_3(aq)$, is electrolyzed using inert electrodes. What is the substance formed at the cathode?

    (A) Na
    (B) N_2
    (C) NO_2
    (D) H_2
    (E) O_2

7. An 1.0 M aqueous solution of zinc iodide, $ZnI_2(aq)$, is electrolyzed using inert electrodes. What is the substance formed at the anode?

    (A) Zn
    (B) I_2
    (C) H_2O
    (D) H_2
    (E) O_2

8. The number of Faradays required to produce 9 g of aluminum, Al, by the electrolysis of molten aluminum oxide, Al_2O_3, is:

    (A) 9 Faradays
    (B) 4 Faradays
    (C) 3 Faradays
    (D) 2 Faradays
    (E) 1 Faraday

9. A current of 3.25 amperes is used to electrolyze a solution of copper II sulfate, $CuSO_4$. How many hours will it take to deposit 12.71 grams of copper, Cu?

    (A) 6.60 hours
    (B) 3.30 hours
    (C) 1.65 hours
    (D) 0.400 hours
    (E) 0.200 hours

10. Two cells, one containing gold nitrate, $Au(NO_3)_3$, and the other copper II sulfate, $CuSO_4$, are connected in series. During the electrolysis, 98.5 grams of gold are deposited in one cell. What weight of copper is deposited in the second cell?

    (A) 21.2 g
    (B) 31.8 g
    (C) 47.6 g
    (D) 63.5 g
    (E) 98.5 g

11. A voltaic cell is set up using the system:
    $Fe/Fe^{2+} \parallel Ag^+/Ag$
    The cathode reaction produces:

    (A) Fe
    (B) Fe^{2+}
    (C) Ag^+
    (D) Ag
    (E) H_2

12. A chemical cell is set up using the system:
    $Zn/Zn^{2+} \parallel Au^{3+}/Au$
    Which is true about movement *through the salt bridge*?

    (A) Zinc ions, Zn^{2+}, move toward the cathode.
    (B) Zinc ions, Zn^{2+}, move towardsthe anode.
    (C) Electrons, e^-, move toward the cathode.
    (D) Gold ions, Au^{3+}, move toward the anode.
    (E) Negative 'spectator' ions move toward the anode.

13. A voltaic cell is set up using the system:
    $Ni/Ni^{2+} \parallel Cu^{2+}/Cu$
    Which reaction would occur at the anode?

    (A) $Ni \Rightarrow Ni^{2+} + 2e^-$
    (B) $Cu \Rightarrow Cu^{2+} + 2e^-$
    (C) $Ni^{2+} + 2e^- \Rightarrow Ni$
    (D) $Cu^{2+} + 2e^- \Rightarrow Cu$
    (E) Both (B) and (C)

14. A voltaic cell employs the reaction:
    $Sn(s) + 2Ag^+(aq) \Rightarrow Sn^{2+}(aq) + 2Ag(s)$
    The voltage produced by this reaction under standard state conditions at 25°C is:

    (A) 1.74 volts
    (B) 1.46 volts
    (C) 0.94 volts
    (D) 0.66 volts
    (E) 0.52 volts

15. The standard voltage, $\mathcal{E}°$, is + 0.68 volts for the voltaic cell:
    $In/In^{3+} \parallel Cu^{2+}/Cu$
    Find the standard oxidation potential for the reaction:
    $In \Rightarrow In^{3+} + 3e^-$

    (A) + 0.34 volts
    (B) -0.34 volts
    (C) +1.02 volts
    (D) -1.02 volts
    (E) 0.00 volts

Multiple-Choice Questions

16. From the Table of Standard Reduction Potentials it can be concluded that:

    (A) Zn^{2+} reacts spontaneously with H_2.
    (B) Ag reacts spontaneously with 1 M hydrochloric acid, HCl.
    (C) Ag reacts spontaneously with 1 M Zn^{2+}.
    (D) Ag reacts spontaneously with 1 M nitric acid, HNO_3.
    (E) Zn^{2+} will liberate H_2 from 1 M HCl.

17. From the Table of Standard Reduction Potentials find a species that will convert Cu^{2+} to Cu but will *not* convert Fe^{2+} to Fe.

    (A) Ag
    (B) H_2
    (C) Sn^{4+}
    (D) I^-
    (E) Zn

18. Which is the strongest reducing agent?

    (A) Ag
    (B) Ag^+
    (C) Fe
    (D) Fe^{2+}
    (E) H_2

19. The potential of the cell:
    $Cd/Cd^{2+} || Ni^{2+}/Ni$
    would be made more positive by increasing the:

    (A) size of the Cd electrode.
    (B) size of the Ni electrode.
    (C) volume of the electrolytes.
    (D) concentration of Cd^{2+}.
    (E) concentration of Ni^{2+}.

20. A voltaic cell has an $\mathcal{E}°$ of 1.56 volts and the overall reaction:
    $$Zn(s) + 2Ag^+(aq) \Leftrightarrow Zn^{2+}(aq) + 2Ag(s)$$

    The $[Zn^{2+}]$ is 0.00010 M and the $[Ag^+]$ is 0.10 M.

    What is the voltage, $\mathcal{E}$, of this cell?

    (A) 1.65 volts
    (B) 1.62 volts
    (C) 1.56 volts
    (D) 1.50 volts
    (E) 1.47 volts

Electrochemistry

Answer Key—Set 8

1.	(E)	6.	(D)	11.	(D)	16.	(D)
2.	(A)	7.	(B)	12.	(A)	17.	(B)
3.	(E)	8.	(E)	13.	(A)	18.	(C)
4.	(C)	9.	(B)	14.	(C)	19.	(E)
5.	(D)	10.	(C)	15.	(A)	20.	(B)

Explanations

1. **(E)**

   Balance O: $ClO_3^- \Rightarrow HClO_2 + H_2O$
   Balance H: $ClO_3^- + 3H^+ \Rightarrow HClO_2 + H_2O$
   Balance charge: $ClO_3^- + 3H^+ + 2e^- \Rightarrow HClO_2 + H_2O$

2. **(A)**

   Balance S: $S_2O_3^{2-} \Rightarrow 2SO_3^{2-}$
   Balance O: $S_2O_3^{2-} + 6OH^- \Rightarrow 2SO_3^{2-} + 3H_2O$
   Balance charge: $S_2O_3^{2-} + 6OH^- \Rightarrow 2SO_3^{2-} + 3H_2O + 4e^-$

3. **(E)** Oxidation is the loss of electrons.
   When the 2 half–reactions are balanced:
   Oxidation: $HSnO_2^- + 2OH^- \Rightarrow HSnO_3^- + 4H_2O + 2e^-$
   Reduction: $Na_2O_2 + 2H_2O + 2e^- \Rightarrow 2Na^+ + 4OH^-$

4. **(C)** The oxidizing agent is reduced in the reaction.
   Method 1:   The oxidation state of nitrogen is being reduced from +5 in NO_3^- to -3 in NH_3.
   Method 2:   Balance the half-reactions to see which is the reduction.
   $NO_3^- + 6H_2O + 8e^- \Rightarrow NH_3 + 9OH^-$

5. **(D)** The oxidation number of sulfur in the sulfite ion, SO_3^{2-}, is +4, an intermediate value for sulfur. Strong oxidizers can oxidize sulfite ion to sulfate ion, SO_4^{2-}. Sulfite can be reduced to either sulfur or to sulfide ion, S^{2-}, by strong reducing agents.

6. **(D)** Na^+ and NO_3^- do not undergo electrolysis in aqueous solution.
   Reduction occurs at the cathode ('Red Cat').
   Reduction: $2H_2O \Rightarrow O_2 + 4H^+ + 4e^-$
   Oxidation: $2H_2O + 2e^- \Rightarrow H_2 + 2OH^-$

7. **(B)** Zn^{2+} and I^- are electrolyzed better than H_2O.
   Oxidation: (anode) $2I^- \Rightarrow I_2 + 2e^-$
   Reduction: (cathode) $Zn^{2+} + 2e^- \Rightarrow Zn$

154 Multiple-Choice Questions

8. **(E)** $n_{Al} = 9 \text{ g} \div 27 \text{ g mol}^{-1} = \frac{1}{3} \text{ mol}$
   $Al^{3+} + 3e^- \Rightarrow Al$
   Faradays $= n_{e^-} = 3n_{Al} = 3 \times \frac{1}{3} = \textbf{1 F}$

9. **(B)** $n_{Cu} = 12.71 \text{ g} \div 63.55 \text{ g mol}^{-1} = 0.2000 \text{ mol}$
   $Cu^{2+} + 2e^- \Rightarrow Cu$
   $n_{e^-} = 2n_{Cu} = 0.4000 \text{ mol}$

   $$\text{Time} = \frac{0.4000 \text{ mol e}^- \times 96,500 \text{ C mol}^{-1}}{3.25 \text{ coul s}^{-1} \times 3600 \text{ s hr}^{-1}} = \textbf{3.30 hr}$$

10. **(C)** $n_{Au} = 98.5 \text{ g} \div 197 \text{ g mol}^{-1} = 0.500 \text{ mol}$
    $Au^{3+} + 3e^- \Rightarrow Au$
    $n_{e^-} = 3n_{Au} = 1.50 \text{ mol}$
    $Cu^{2+} + 2e^- \Rightarrow Cu$
    $n_{Cu} = \frac{1}{2}n_{e^-} = 0.750 \text{ mol}$
    $wt_{Cu} = 0.750 \text{ mol} \times 63.55 \text{ g mol}^{-1} = \textbf{47.6 g}$

11. **(D)** Reduction occurs at the cathode ('Red Cat').
    Cathode: $Ag^+ + e^- \Rightarrow \textbf{Ag}$

12. **(A)** Cathode: $Au^{3+} + 3e^- \Rightarrow Au$
    The [Au^{3+}] decreases in the cathode half-reaction.
    There must be a migration of (+) ions ($\textbf{Zn}^{2+}$) to the cathode and (−) ions to the anode to balance the charge in each cell.
    Electrons can only move in the external circuit.

13. **(A)** **Oxidation occurs at the anode ('An Ox').**
    Anode: $Ni \Rightarrow Ni^{2+} + 2e^-$

14. **(C)**

		$\mathcal{E}°$
Anode:	$Sn \Rightarrow Sn^{2+} + 2e^-$	+ 0.14 volts
Cathode:	$2Ag^+ + 2e^- \Rightarrow 2Ag$	+ 0.80 volts
Overall:	$Sn + 2Ag^+ \Rightarrow Sn^{2+} + 2Ag$	**+ 0.94 volts**

15. **(A)**

		$\mathcal{E}°$
ANODE:	$In \Rightarrow In^{3+} + 3e^-$	? volts
CATHODE:	$Cu^{2+} + 2e^- \Rightarrow Cu$	+ 0.34 volts
OVERALL:	$In + Cu^{2+} \Rightarrow In^{2+} + Cu$	+ 0.68 volts

$\mathcal{E}°_{In}$ = **0.34 volts**

16. **(D)** A positive net cell potential is associated with spontaneous reactions. Nitric acid is a strong oxidizing agent.

Oxidation:	$3Ag \Rightarrow 3Ag^+ + 3e^-$	– 0.80 volts
Reduction:	$NO_3^- + 4H^+ + 3e^- \Rightarrow NO + 2H_2O$	+ 0.96 volts
		+ 0.16 volts

17. **(B)** H_2 is between Fe^{2+} and Cu^{2+} in the table. It will give a positive net cell potential with Cu^{2+}, but not Fe^{2+}.

$Fe^{2+} + 2e^- \Rightarrow Fe$	– 0.45 volts
$2H^+ + 2e^- \Rightarrow H_2$	+ 0.00 volts
$Cu^{2+} + 2e^- \Rightarrow Cu$	+ 0.34 volts

18. **(C)** The strongest reducing agent is the most easily oxidized. The highest *oxidation* potential of those given is that of **iron, Fe**, whose $\mathcal{E}°$ is + 0.45 volts.

19. **(E)** The net cell reaction is:
$$Cd + Ni^{2+} \Leftrightarrow Cd^{2+} + Ni$$
A voltaic or chemical cell at equilibrium is a *dead battery*. **Increasing $[Ni^{2+}]$** (a reactant) and decreasing $[Cd^{2+}]$ (a product) will, from LeChatelier's Principle, shift the concentrations further away from equilibrium conditions. These concentration changes will increase the net cell potential—cell voltage is a measure of how far the system is from equilibrium.

Changing the electrode size or the electrolyte volume increases the amount, but has no effect on the voltage. The amount of a solid or a liquid will not change its concentration and will not affect the reaction rate of an equilibrium reaction.

20. **(B)** $\mathcal{E} = \mathcal{E}° - \dfrac{0.0591}{n} \log \dfrac{[Zn^{2+}]}{[Ag^+]^2}$

$\mathcal{E} = 1.56 - \dfrac{0.059}{2} \log \dfrac{[0.00010]}{[0.10]^2}$

$\mathcal{E}$ = **1.62 volts**

Part III

Free Response Section

About Section II

Free Response Problems

Section II of the AP examination generally constitutes 55% of the final score. Between 75 and 90 minutes is allotted for completion, with 90 minutes being the most usual time allowance. This section is a distinctive and important feature of the examination.

Parts A and B are problems that permit the candidate to demonstrate an ability to apply chemical principles to the solution of college-level problems. Part A is worth 20 points and has always been a question about 'Equilibrium'. Part B is also worth 20 points and permits the choice of one problem (from two given) dealing with 'Electrochemistry', 'Kinetics', 'Stoichiometry', or 'Thermodynamics'. No more than 20-25 minutes should be spent on each part.

This table is a history of the types of problems given for each of the past years. It should be understood that no prediction can be made for any specific test. Section II questions are released each year. Collections of Section II questions, answers and explanations are published by the Educational Testing Service.

Year	Part A (Mandatory) Equilibrium	Part B (Choose 1 of 2 questions) Electrochemistry, Kinetics, Stoichiometry, Thermo	
1976	Acid-Base $[H_3O^+]$, K_a, K_{sp}	Electrochemistry $\Delta G°$, $E°$, Faraday	Stoichiometry Molality, Raoult Colligative Properties
1977	Acid-Base $[H_3O^+]$, buffers, pH	Kinetics K_{eq}, Rate	Thermo $\Delta G°$, $\Delta S°$, K_{eq}
1978	Acid-Base Titration	Electrochemistry $\Delta G°$, $E°$, Stoichiometry	Thermo $\Delta H°$, K, LeChatelier
1979	Solubility K_{sp}	Stoichiometry Solutions, molarity	Thermo $\Delta G°$, $\Delta S°$
1980	Acid-Base $[H_3O^+]$, K_b, K_{sp}	Electrochemistry $E°$, Nernst	Stoichiometry Colligative, empirical, gas laws
1981	Gaseous K_p, gas law	Kinetic Rate and k	Stoichiometry Balance, titration
1982	Acid-Base $[H_3O^+]$, pH, buffers	Electrochemistry Electrolysis, Faraday	Stoichiometry Empirical, percent composition
1983	Gaseous Law, Dalton K_p, K_c	Stoichiometry Solutions, titration molar mass	Thermo $\Delta G°$, K, $\Delta H°$, $\Delta S°$
1984	Acid-Base $[OH^-]$, K_b, K_a, solubility	Kinetics Rate law, rate constant, mechanisms	Thermo $\Delta G°$, $\Delta H°$, $\Delta S°$

Year	Part A (Mandatory) *Equilibrium*	Part B (Choose 1 of 2 questions) *Electrochemistry, Kinetics, Stoichiometry, Thermo*	
1985	*Solubility* Selective precipitation	*Electrochemistry* Balance, Faraday, $E°$, $\Delta G°$	*Stoichiometry* Empirical, molecular, colligative, MW
1986	*Acid-Base* K_a, pH, buffers	*Electrochemistry* Faraday, balancing, redox, titration	*Stoichiometry* Gas law, molar mass, empirical formula
1987	*Acid-Base* $[OH^-]$, pH, buffers, K_{sp}	*Kinetics* Rate law, stoichiometry	*Stoichiometry* Solutions, titration, percent composition
1988	*Gaseous* Gas law, K_c, K_p	*Electrochemistry* Electrochemical cells, K from $E°$	*Thermo* Hess' Law, $\Delta G°$, $\Delta H°$, $\Delta S°$, K
1989	*Acid-Base* Molar mass, K_a, buffers, pH	*Electrochemistry* Titration, stoichiometry, redox reaction, Faraday	*Thermo* $\Delta G°$, $\Delta H°$, $\Delta S°$, T, vapor pressure
1990	*Solubility* K_{sp}, pH, ppt formation: Q vs K_{sp}	*Stoichiometry* Partial pressure, gas law, vapor pressure	*Thermo* $\Delta H°$ from bond energy $\Delta G°$, $\Delta S°$, K_{eq}, LeChatelier
1991	*Acid-Base* K_a, $[H^+]$, % ionization buffers	*Stoichiometry* Combustion, colligative, molecular formula	*Kinetics* Rate law, rate constant, reaction mechanism
1992	*Gaseous* Gas law, stoichiometry, K_p	*Electrochemistry* Electrolysis, voltaic cells, $\Delta G°$ from $E°$, Nernst	*Thermo* K from $\Delta G°$, entropy ($\Delta S°$) calculations & questions

The Part A and B problems usually have three or four subparts. Subparts (a) and (b) are standard questions that have usually been practiced frequently during the year in the AP course. Subparts (c) and (d) are a little more elegant—and difficult—since they *seem* to be unique. The proper application of reviewed principles will generally yield a solution to all of the subparts.

The correct strategy for candidates working on this section is to make sure to:

- show a logical series of steps that can lead to an answer.

- acknowledge that only one-third credit is given for a right answer without supporting steps.

- appreciate that partial credit is given for each correct step in the attempt to answer the question, even if the question is not fully answered.

- realize that some problems have several valid approaches to the correct answer.

- make use of the fact that an incorrect answer from one part will not be penalized when used in other parts of the question.

- consistently show units in equations and when labeling answers.

- use significant figures correctly.

Chapter 1

Part A Sample Problems

1. **Gaseous Equilibrium**

   Solid ammonium chloride is placed in an evacuated vessel and heated until it decomposes.
   $$NH_4Cl(s) \Leftrightarrow NH_3(g) + HCl(g)$$
   (a) Equilibrium is reached at a temperature of 202°C, and the total pressure inside the vessel is 4.40 atm.
   What is the numerical value of K_p at 202°C?

   (b) Ammonia gas, $NH_3(g)$, is added to the vessel, which is kept at 202°C. The system returns to equilibrium, and the partial pressure of the NH_3 is found to be three-times that of the HCl.
   What are the partial pressures of NH_3 and HCl?

   (c) To an empty 1.00 liter vessel is added 0.130 mole each of NH_3 and HCl at 202°C. How many moles of solid NH_4Cl will be formed at this temperature?

   (d) Explain the effect on the equilibrium constant of raising the temperature to 600 K.

2. **Solubility Equilibrium**

   A saturated solution of magnesium hydroxide, $Mg(OH)_2$, has a magnesium ion concentration of 1.65×10^{-4} M at 25°C.

   (a) What is the value of the solubility product constant, K_{sp}, at 25°C.

   (b) What is the molar solubility of $Mg(OH)_2$ in 0.10 M $Mg(NO_3)_2$ solution at 25°C.

   (c) To 350. mL of a 0.150 M $Mg(NO_3)_2$ solution, 150. mL of a 0.500 M NaOH solution is added. What are the $[Mg^{2+}]$ and $[OH^-]$ in the resulting solution at 25°C?

3. Acid-Base Equilibrium

In water, acetylsalicylic acid (aspirin), $HC_9H_7O_4$, is a weak acid that has an equilibrium constant, K_a, equal to 2.75×10^{-5} at 25°C. A 0.400 liter sample of a 0.100 M solution of the acid is prepared.

(a) What are the equilibrium concentrations of $C_9H_7O_4^-$, H_3O^+, OH^- and what is the pH of the solution at 25°C?

(b) What is the K_b for the reaction:
$$C_9H_7O_4^- + H_2O \Leftrightarrow HC_9H_7O_4 + OH^-$$

(c) To 0.2000 liter of the solution, 3.03 grams of sodium acetysalicylate, $NaC_9H_7O_4$ (202 g mol^{-1}), is added. The salt dissolves completely and the volume of the solution remains unchanged.
Calculate the pH of the resulting solution at 25°C.

(d) To the remaining 0.200 liter of the original solution, 0.100 liter of 0.100 M NaOH solution is added.
Calculate the $[OH^-]$ for the resulting solution at 25°C.

4. Acid-Base Titration

A 0.4157 gram sample is dissolved in 50.00 mL of water in a titration experiment to determine the molar mass and ionization constant, K_a, of lactic acid. The lactic acid solution is titrated with a standardized 0.1085 M NaOH solution at a temperature of 25°C. The equivalence point is reached after 42.53 mL of base is added. Lactic acid is a monoprotic acid that may be represented as HLac.

(a) Calculate the molar mass of lactic acid from these data.

(b) The pH of the solution is 2.81 after 15.00 mL of base is added. Calculate the ionization constant, K_a, of lactic acid.

(c) Calculate the equilibrium constant, K_b, of lactate ion, Lac$^-$, reacting as a base with water.

(d) Calculate the pH of the solution at the equivalence point of the titration.

Sample Answers and Explanations

1. **Gaseous Equilibrium**

   (a) Answer: $K_p = 4.84$ atm^2

	NH$_4$Cl(s)	⇔ NH$_3$(g)	+ HCl(g)
Equilibrium Partial Pressure	0	x	x

   $P_{total} = 0 + x + x = 4.40$ atm.
   $x = P_{NH_3} = P_{HCl} = 2.20$ atm.
   $K_p = (P_{NH_3}) \times (P_{HCl}) = (2.20 \text{ atm})^2 = 4.84$ atm^2

   (b) Answers: $P_{HCl} = 1.27$ atm
   $P_{NH_3} = 3.81$ atm

   $K_p = (P_{NH_3}) \times (P_{HCl}) = (3P_{HCl}) \times (P_{HCl}) = 4.84$ atm^2
   $P_{HCl} = 1.27$ atm
   $P_{NH_3} = 3P_{HCl} = 3 \times 1.27 = 3.81$ atm

   (c) Answer: $n_{NH_4Cl} = 0.0736$ mol

   Since moles in a 1.00 L volume are given (molarity), K_c should be handier to use than K_p.

   $$K_c = K_p \left(\frac{1}{RT}\right)^{\Delta n}$$

   $$K_c = 4.84 \text{atm}^2 \left(\frac{1}{0.0821 \text{ L atm mol}^{-1} \text{ K}^{-1} \times 475 \text{ K}}\right)^2$$

   $K_c = 0.00318$ M^2

	NH$_4$Cl(s)	⇔ NH$_3$(g)	+ HCl(g)
Start	0	0.130	0.130
Δ	+x	-x	-x
Equilibrium	x	0.130 - x	0.130 - x

   $K_c = [NH_3][HCl] = [0.130 - x]^2 = 0.00318$ M^2
   $x = n_{NH_4Cl} = 0.0736$ mol

   (d) The equilibrium constant will be larger at 600 K.

   The forward reaction is endothermic. An increase in temperature favors the endothermic (heat consuming) reaction leading to an increase in the partial pressure of the products.

2. **Solubility Equilibrium**

   (a) **Answer:** $K_{sp} = 1.80 \times 10^{-11}$

   $[Mg^{2+}] = 1.65 \times 10^{-4}$ M
   $[OH^-] = 2 \times [Mg^{2+}] = 3.30 \times 10^{-4}$ M
   $K_{sp} = [Mg^{2+}][OH^-]^2 = [1.65 \times 10^{-4}][3.30 \times 10^{-4}]^2$
   $K_{sp} = 1.80 \times 10^{-11}$

   (b) **Answer:** x = molar solubility = 6.71×10^{-6} M

   Let x equal the molar solubility of $Mg(OH)_2$.

	$Mg(OH)_2(s)$	$\Leftrightarrow Mg^{2+}(aq)$	$+ 2OH^-(aq)$
Start	0	0.10 M	0 M
Δ	-x	+x	+2x
Finish	Some - x	0.10	2x
Note: x is small compared to 0.10, and is ignored.			

   $K_{sp} = [0.10][2x]^2 = 1.80 \times 10^{-11}$
   x = molar solubility = 6.71×10^{-6} M

   (c) **Answers:** $[Mg^{2+}] = 0.030$ M
   $[OH^-] = x = 2.45 \times 10^{-5}$ M

   $[Mg^{2+}] = \dfrac{350.\text{ mL} \times 0.150 \text{ mol L}^{-1}}{500.\text{ mL}} = 0.105$ M
   $[OH^-] = \dfrac{150.\text{ mL} \times 0.500 \text{ mol L}^{-1}}{500.\text{ mL}} = 0.150$ M

   Let x = amount of OH^- formed as a new equilibrium is established ('bounce back').

	$Mg(OH)_2(s)$	$\Leftrightarrow Mg^{2+}(aq)$	$+ 2OH^-(aq)$
Start	0	0.105 M	0.150 M
Δ_1	+0.075	-0.075	-0.150
Precipitate	Some	0.030	0
Δ_2	-x/2	+x/2	+x
Note: x/2 is small compared to 0.030, and is ignored.			
Finish	Some - x/2	0.030	x

   $K_{sp} = [Mg^{2+}][OH^-]^2 = [0.030][x]^2 = 1.80 \times 10^{-11}$
   $[Mg^{2+}] = 0.030$ M
   $[OH^-] = x = 2.45 \times 10^{-5}$ M

3. **Acid-Base Equilibrium**

   (a) Answers: $[H_3O^+]$ and $[C_9H_7O_4^-]$ = 1.66×10^{-3} M

   pH = 2.780; $[OH^-]$ = 6.02×10^{-12} M

	$HC_9H_7O_4(aq)$	$\Leftrightarrow$ $H^+(aq)$	$C_9H_7O_4^-(aq)$
Start	0.100 M	0.0 M	0.0 M
Δ	-x	+x	+x
Finish	0.100 - x	x	x
Note: x is small compared to 0.100, and is ignored.			

   $$K_a = \frac{[H^+][C_9H_7O_4^-]}{[HC_9H_7O_4]} = \frac{[x][x]}{[0.100]} = 2.75 \times 10^{-5}$$

   x = $[H^+(aq)]$ = $[H_3O^+]$ = $[C_9H_7O_4^-]$ = 1.66×10^{-3} M

   pH = $-\log[H^+]$ = 2.780

   K_w = $[H^+][OH^-]$ = $[1.66 \times 10^{-3}][x]$ = 1.0×10^{-14}

   x = $[OH^-]$ = 6.02×10^{-12} M

   (b) Answer: K_b = 3.6×10^{-10}

   $K_a \times K_b$ = K_w = 1.0×10^{-14}

   $(2.75 \times 10^{-5}) \times K_b$ = 1.0×10^{-14}

   K_b = 3.6×10^{-10}

   (c) Answer: pH = 4.436

   $n_{C_9H_7O_4^-}$ = 3.03 g ÷ 202 g mol^{-1} = 0.0150 mol

   molarity = 0.0150 mol ÷ 0.200 L = 0.0750 M

   $$pH = pK_a + \log\frac{[C_9H_7O_4^-]}{[HC_9H_7O_4]} = 4.561 + \log\frac{[0.0750]}{[0.100]}$$

   pH = 4.436

(d) Answer: $[OH^-] = 3.6 \times 10^{-10}$ M

$$[OH^-] = \frac{0.100 \text{ L} \times 0.100 \text{ mol L}^{-1}}{0.300 \text{ L}} = 0.0333 \text{ M}$$

$$[HC_9H_7O_4] = \frac{0.200 \text{ L} \times 0.100 \text{ mol L}^{-1}}{0.300 \text{ L}} = 0.0667 \text{ M}$$

	$C_9H_7O_4^-$(aq)	+ H_2O	$\Leftrightarrow$ $HC_9H_7O_4$(aq)	+ OH^-(aq)
Start	0	Some	0.0667 M	0.0333 M
Δ_1	+0.0333	+0.0333	−0.0333	−0.0333
Finish$_1$	0.0333	Some	0.0334	0.00
Δ_2	−x	−x	+x	+x
Note: x is small compared to 0.0333 and is ignored.				
Finish$_2$	0.0333	Some	0.0334	x

$$K_b = \frac{[HC_9H_7O_4][OH^-]}{[C_9H_7O_4^-]} = \frac{[0.0334]x}{[0.0333]} = 3.6 \times 10^{-10}$$

$x = [OH^-] = 3.6 \times 10^{-10}$ M

4. **Acid-Base Titration**

(a) Answer: MM = 90.09 g mol^{-1}

At the *equivalence* point:

$n_{HLtc} = n_{OH^-} = M_{OH^-} \cdot V_{OH^-} = 0.1085 \times 0.04253 = 4.614 \times 10^{-3}$ mol

$$MM = \frac{0.4157 \text{ g}}{4.614 \times 10^{-3}} = 90.09 \text{ g mol}^{-1}$$

(b) Answer: $K_a = 8.5 \times 10^{-4}$

$$pH = pK_a + \log \frac{[A^-]}{[HA]}$$

There are 65.00 mL of solution after the NaOH is added to the acid solution.

$$[HLac]_{start} = \frac{4.614 \times 10^{-3} \text{ mol}}{0.06500 \text{ L}} = 0.07098 \text{ M}$$

$$[OH^-]_{start} = \frac{0.1085 \times 15.00}{65.00} = 0.02504 \text{ M}$$

$Lac^- + H_2O \Leftrightarrow HLac + OH$

$[Lac^-] = [HLac]_{start} = 0.02504$ M

$[HLac] = [HLac]_{start} - [OH^-]_{added} = 0.07098 - 0.02504 = 0.04594$ M

$$pK_a = pH - \log \frac{[Ltc^-]}{[HLtc]} = 2.81 - \log \frac{[0.02504]}{[0.04594]} = 3.07$$

$K_a = 8.5 \times 10^{-4}$

(c) **Answer: $K_b = 1.2 \times 10^{-11}$**

$$K_b = \frac{1.0 \times 10^{-14}}{K_a} = \frac{1.0 \times 10^{-14}}{8.5 \times 10^{-4}} = 1.2 \times 10^{-11}$$

(d) **Answer: pH = 7.89**

There are 92.53 mL of solution at the equivalence point.

$$[HLac]_{start} = \frac{4.614 \times 10^{-3} \text{ mol}}{0.09253 \text{ L}} = 0.04986 \text{ M}$$

$$[OH^-]_{start} = \frac{0.1085 \times 42.53}{92.53} = 0.04987 \text{ M}$$

Equation	Lac^-	$+ H_2O$	$\Leftrightarrow HLac$	$+ OH^-$
Start	0	Some	0.0499 M	0.0499 M
Δ_1	+0.0499	+0.0499	-0.0499	-0.0499
Finish$_1$	0.0499	Some	0.00	0.00
Δ_2	-x	-x	+x	+x
Finish$_2$	0.0499 + x	Some	x	x
Note: x is small compared to 0.0499, and is ignored.				
Finish$_2$	0.0499	Some	x	x

$$\frac{x^2}{0.0499} = 1.2 \times 10^{-11}$$

$$x = [OH^-] = 7.7 \times 10^{-7}$$

$$[H^+] = \frac{1.0 \times 10^{-14}}{[OH^-]} = \frac{1.0 \times 10^{-14}}{7.7 \times 10^{-7}} = 1.3 \times 10^{-8}$$

$$pH = -\log[1.3 \times 10^{-8}] = 7.89$$

Chapter 2

Part B Sample Problems

1. **Electrochemistry—Electrolytic Cells**

    An acidic solution contains lead(II) ions, Pb^{2+}, and an anion that is not electrolyzed. Electrolysis of the solution between lead electrodes produces lead IV oxide, PbO_2, at the anode and lead, Pb, at the cathode.

    (a) Write the balanced equation for the anode reaction.

    (b) A current of 1.75 amperes is applied to the system for 20.0 minutes. How many grams of Pb will be plated out at the cathode?

    (c) In series with the cell described above is another cell consisting of a solution of chromium(III) nitrate, $Cr(NO_3)_3$, and two platinum electrodes. How many grams of chromium, Cr, will be plated out when 49.5 grams of lead, Pb, are produced in the other cell?

2. **Electrochemistry—Voltaic Cells**

    A voltaic cell consists of a copper electrode in a solution of copper(II) ions, $Cu^{2+}(aq)$, and a paladium electrode in a solution of paladium(II) ions, $Pd^{2+}(aq)$.
    The copper electrode is the anode. The reduction potential for copper(II) ions is +0.342 volts.

    (a) Write the half-cell equation for the reaction that occurs at the cathode.

    (b) The standard cell voltage, $\mathcal{E}°$, is 0.609 volts. What is the reduction potential for the Pd^{2+}/Pd half-reaction?

    (c) What is the value of the equilibrium constant, K, for this reaction at 25°C?

    (d) Consider a nonstandard cell of $Cu/Cu^{2+} \| Pd^{2+}/Pd$. The molarity of the copper(II) ion, Cu^{2+}, is 3.00 M and the molarity of the paladium(II) ion, $[Pd^{2+}]$, is 0.0500 M. What is the cell voltage?

3. **Kinetics**

The data were obtained for the initial rate of the reaction:
$$2NO(g) + O_2(g) \Rightarrow 2NO_2(g)$$

Data	[NO]	[O_2]	Initial rate; mol L^{-1} s^{-1}
Run 1	0.20 M	0.20 M	4.64×10^{-8}
Run 2	0.20	0.40	9.28×10^{-8}
Run 3	0.40	0.40	3.71×10^{-7}
Run 4	0.10	0.10	5.80×10^{-9}

(a) What is the rate law equation for the reaction?

(b) What is the value of the specific rate law constant, including units?

(c) The following mechanism has been suggested:

1. $NO(g) + O_2(g) \Leftrightarrow NO_3(g)$ (Equilibrium)
2. $NO_3(g) + NO(g) \Rightarrow 2NO_2(g)$ (Slow)

Show that this mechanism leads to the observed rate equation.

(d) The reaction rate is increased 5-fold when the temperature is increased from 1400 K to 1500 K. What is the energy of activation for the reaction?

4. **Stoichiometry—Empirical Formulas and Colligative Properties**

Ascorbic acid (Vitamin C) is a compound known to contain carbon, hydrogen, and oxygen.

(a) Combustion analysis of a 1.000 gram sample of ascorbic acid yields 1.500 g of CO_2 and 0.409 g of H_2O. What is the empirical formula of this acid?

(b) When another 1.00 g sample of ascorbic acid is dissolved in 10.0 g of water the freezing point of the solution is -1.06°C. At this temperature the dissociation of ascorbic acid is negligible. The freezing point constant, K_f, of water is 1.86 °C molal^{-1}. What is the molar mass of ascorbic acid?

(c) What is the molecular formula of ascorbic acid?

5. **Stoichiometry—Standard Solutions**

   A sample of iron(II) ion, Fe^{2+}, is standardized with an acid solution of dichromate ion, $Cr_2O_7^{2-}$.

   (a) Write the balanced equation for the titration reaction, shown in an unbalanced 'skeleton' form.
   $$Fe^{2+} + Cr_2O_7^{2-} \Rightarrow Fe^{3+} + Cr^{3+}$$

   (b) A standard is prepared by dissolving 0.1176 g of potassium dichromate, $K_2Cr_2O_7$ (molar mass 294.1), in 100.0 mL of a water solution containing hydrochloric acid. The standard required 28.6 mL of Fe^{2+} solution to completely reduce the $Cr_2O_7^{2-}$. What is the molarity of the Fe^{2+} solution?

   (c) It requires 26.9 mL of the standardized Fe^{2+} solution to neutralize 25.0 mL of an acidified permanganate, MnO_4^-, solution. What is the molarity of the MnO_4^- solution?
   The balanced equation for the reaction is:
   $$5Fe^{2+} + MnO_4^- + 8H^+ \Rightarrow 5Fe^{3+} + Mn^{2+} + 4H_2O$$

6. **Thermodynamics**

   Methane, CH_4, and chlorine, Cl_2, are reacted to form carbon tetrachloride, CCl_4, and hydrogen chloride, HCl at 25°C. The heat of the reaction, $\Delta H°$, is -429.8 kJ mol^{-1} of CH_4, reacted. The equation is:
   $$CH_4(g) + 2Cl_2(g) \Rightarrow CCl_4(l) + 4HCl(g)$$

Substance	Heat of Formation, $\Delta H_f°$, kJ mol^{-1}	Absolute Entropy, $S°$, J mol^{-1} K^{-1}
$C_{graphite}(s)$	0	5.740
$CH_4(g)$	-74.86	186.2
$CCl_4(l)$	?	216.4
$Cl_2(g)$	0	223.0
$HCl(g)$	-92.31	186.8

   (a) What is the standard heat of formation, $\Delta H_f°$, for carbon tetrachloride at 25°C?

   (b) Calculate the standard entropy change, $\Delta S°$, for the reaction at 25°C.

   (c) Theoretically, carbon tetrachloride can be formed by the reaction:
   $$C_{graphite}(s) + 2Cl_2(g) \Rightarrow CCl_4(l)$$
   Calculate the standard free energy of formation, $\Delta G_f°$, for carbon tetrachloride.

   (d) Calculate the value of the equilibrium constant, K, for the reaction in Part (c) at 25°C.

Sample Answers and Explanations

1. **Electrochemistry—Electrolytic Cells**

    (a) **Answer:** $Pb^{2+} + 2H_2O \Rightarrow PbO_2 + 4H^+ + 2e^-$
    The oxidation reaction occurs at the anode.

    (b) **Answer: 2.25 grams**

    $$\text{Faradays} = \frac{1.75 \text{ C}}{1 \text{ sec}} \times 1200 \text{ sec} \times \frac{1 \text{ mol e}^-}{96500 \text{ C}}$$

    Faradays = 0.0218 mol e$^-$
    Cathode (Reduction): $Pb^{2+} + 2e^- \Rightarrow Pb$

    $$g_{Pb} = 0.0218 \text{ mol e}^- \times \frac{1 \text{ mol Pb}}{2 \text{ mol e}^-} \times \frac{207 \text{ g Pb}}{1 \text{ mol Pb}} = 2.25 g$$

    (c) **Answer: 8.29 grams**

    $$\text{Faradays} = 49.5 \text{ g Pb} \times \frac{1 \text{ mol Pb}}{207 \text{ g Pb}} \times \frac{2 \text{ mol e}^-}{1 \text{ mol Pb}}$$

    Faradays = 0.478 mol e$^-$
    Cathode (Reduction): $Cr^{3+} + 3e^- \Rightarrow Cr$

    $$g_{Cr} = 0.478 \text{ mol e}^- \times \frac{1 \text{ mol Cr}}{3 \text{ mol e}^-} \times \frac{52.0 \text{ g Cr}}{1 \text{ mol Cr}} = 8.29 g$$

2. **Electrochemistry—Voltaic Cells**

    (a) **Answer:** $Pd^{2+} + 2e^- \Rightarrow Pd$
    Cathode (Reduction): $Pd^{2+} + 2e^- \Rightarrow Pd$
    Anode (Oxidation): $Cu \Rightarrow Cu^{2+} + 2e^-$

    (b) **Answer:** $\varepsilon° = 0.951$ volts

			$\varepsilon°$
Anode	(Oxidation):	$Cu \Rightarrow Cu^{2+} + 2e^-$	-0.342 volts
Cathode	(Reduction):	$Pd^{2+} + 2e^- \Rightarrow Pd$	x
Overall		$Cu + Pd^{2+} \Rightarrow Pd + Cu^{2+}$	+0.609

    x - 0.342 volts = 0.609 volts $\varepsilon° = x = 0.951$ volts

    (c) **Answer:** $K = 4 \times 10^{20}$

    $$\varepsilon° = \frac{0.0591}{n} \log K \quad \text{(Nernst equation; } \varepsilon = 0 \text{ at equilibrium)}$$

    $$0.609 \text{ volts} = \frac{0.0591}{2 \text{ mol}} \log K$$

    $K = 4 \times 10^{20}$

    (d) **Answer:** $\varepsilon = 0.557$ volts

    $$\varepsilon = \varepsilon° - \frac{0.0591}{n} \log \frac{[Cu^{2+}]}{[Pd^{2+}]} = 0.609 - \frac{0.0591}{2} \log \frac{[3.00]}{[0.0500]} = 0.557 \text{ volts}$$

3. Kinetics

(a) **Answer: Rate = $k[NO]^2[O_2]$**

Rate = $k[NO]^m[O_2]^n$

$$\frac{\text{Run 2}}{\text{Run 1}} = \frac{9.28 \times 10^{-8}}{4.64 \times 10^{-8}} = \left[\frac{0.20}{0.20}\right]^m \times \left[\frac{0.40}{0.20}\right]^n$$

$2 = [2]^n \qquad n = 1$

$$\frac{\text{Run 3}}{\text{Run 2}} = \frac{3.71 \times 10^{-7}}{9.28 \times 10^{-8}} = \left[\frac{0.40}{0.20}\right]^m \times \left[\frac{0.40}{0.40}\right]^n$$

$4 = [2]^m \qquad m = 2$

(b) **Answer: $k = 5.8 \times 10^{-6}\ M^{-2}\ s^{-1}$**

From Run 4 (or any convenient run):

Rate = 5.80×10^{-9} mol L^{-1} s^{-1} = $k[0.10\ M]^2[0.10\ M]$
$k = 5.80 \times 10^{-6}\ M^{-2}\ s^{-1}$

(c) **Answer: Rate = $K_1 k_2 [NO]^2[O_2] = k[NO]^2[O_2]$**

Rate = $k_2[NO_3][NO]$

$K_1 = \dfrac{[NO_3]}{[NO][O_2]}$

$[NO_3] = K_1[NO][O_2]$
Rate = $K_1 k_2 [NO]^2 [O_2]$

(d) **Answer: 280 kJ**

$$\ln\frac{k_2}{k_1} = -\frac{E_a}{R}\left[\frac{1}{T_2} - \frac{1}{T_1}\right]$$

$$\ln\frac{5}{1} = -\frac{E_a}{8.314\ J\ mol^{-1}\ K^{-1}}\left[\frac{1}{1500} - \frac{1}{1400}\right]$$

$E_a = 280{,}000\ J = 280\ kJ$

4. Stoichiometry—Empirical Formulas and Colligative Properties

(a) **Answer: $C_3H_4O_3$**

$$n_C = n_{CO_2} = \frac{1.500\ g}{44.0\ g\ mol^{-1}} = 0.0341\ mol$$

$$n_H = 2 \times n_{H_2O} = \frac{(2 \times 0.409\ g)}{18.0\ g\ mol^{-1}} = 0.0454\ mol$$

	wt	mol	ratio	empirical
Carbon, C	0.409 g	0.0341 mol	1.00	3
Hydrogen, H	0.045 g	0.0454 mol	1.33	4
Oxygen, O	0.546 g	0.0341 mol	1.00	3
Total	1.000 g	Empirical formula is $C_3H_4O_3$		

(b) **Answer: 175 g mol^{-1}**

$$\frac{wt_{ascorbic}}{1.0\,kg_{H_2O}} = \frac{1.00\,g_{ascorbic}}{10.0\,g_{H_2O}} \times \frac{1000\,g_{H_2O}}{1\,kg_{H_2O}} = 100.\,g\,kg_{H_2O}^{-1}$$

$$m = \frac{\Delta t}{K_f} = \frac{-1.06\,°C}{-1.86\,°C\,m^{-1}} = 0.570\,mol\,kg^{-1}$$

$$\text{Molar mass} = \frac{100.\,g\,kg^{-1}}{0.570\,mol\,kg^{-1}} = 175\,g\,mol^{-1}$$

(c) **Answer: $C_6H_8O_6$**

The empirical mass of $C_3H_4O_3$ is 88 g mol^{-1}

$$\text{\# of empirical formulas} = \frac{\text{molar mass}}{\text{empirical mass}} = \frac{176\,g\,mol^{-1}}{88\,g\,mol^{-1}} = 2.0$$

Molecular formula = 2 × $C_3H_4O_3$ = $C_6H_8O_6$

5. **Stoichiometry—Standard Solutions**

(a) **Answer: $6Fe^{2+} + Cr_2O_7^{2-} + 14H^+ \Rightarrow 6Fe^{3+} + 2Cr^{3+} + 7H_2O$**

Step	Action	Procedure
1	Half-reactions:	$Fe^{2+} \Rightarrow Fe^{3+}$ $Cr_2O_7^{2-} \Rightarrow Cr^{3+}$
2	Balance atoms, except O and H:	$Fe^{2+} \Rightarrow Fe^{3+}$ $Cr_2O_7^{2-} \Rightarrow 2Cr^{3+}$
3	Balance oxygen with H$_2$O:	$Fe^{2+} \Rightarrow Fe^{3+}$ $Cr_2O_7^{2-} \Rightarrow 2Cr^{3+} + 7H_2O$
4	Balance hydrogen with H$^+$:	$Fe^{2+} \Rightarrow Fe^{3+}$ $Cr_2O_7^{2-} + 14H^+ \Rightarrow 2Cr^{3+} + 7H_2$
5	Balance the charge with electrons:	$Fe^{2+} \Rightarrow Fe^{3+} + e^-$ $Cr_2O_7^{2-} + 14H^+ + 6e^- \Rightarrow 2Cr^{3+} + 7H_2O$
6	Balance the electrons:	$6Fe^{2+} \Rightarrow 6Fe^{3+} + 6e^-$ $Cr_2O_7^{2-} + 14H^+ + 6e^- \Rightarrow 2Cr^{3+} + 7H_2O$
7	Combine and simplify: $6Fe^{2+} + Cr_2O_7^{2-} + 14H^+ \Rightarrow 6Fe^{3+} + 2Cr^{3+} + 7H_2O$	

(b) **Answer: [Fe^{2+}] = 0.0839 mol L^{-1}**

$$n_{Cr_2O_7^{2-}} = \frac{0.1176\,g}{294.1\,g\,mol^{-1}} = 0.0003999\,mol$$

$$n_{Fe^{2+}} = 6n_{Cr_2O_7^{2-}} = 0.002399\,mol$$

$$[Fe^{2+}] = \frac{0.002399\,mol}{0.0286\,L} = 0.0839\,M$$

(c) **Answer: $M_{MnO_4} = 0.0181$ mol L^{-1}**

$$n_{permanganate} = \tfrac{1}{5} n_{Fe}$$
$$M_{MnO_4} \times V_{MnO_4} = \tfrac{1}{5} \times M_{Fe} \times V_{Fe}$$
$$M_{MnO_4^{1-}} = \frac{1}{5} \frac{(0.0839 \text{ M} \times 26.9 \text{ mL})}{25.0 \text{ mL}}$$
$$M_{MnO_4} = 0.0181 \text{ mol L}^{-1}$$

6. **Thermodynamics**

(a) **Answer:** $\Delta H_f° = -135.4$ kJ mol^{-1}

$\Delta H° = \Sigma \Delta H_f°(\text{Products}) - \Sigma \Delta H_f°(\text{Reactants})$

-429.8 kJ $= (\Delta H_f° + (4 \times -92.31 \text{ kJ})) - (-74.86 + 0)$

$\Delta H_f° = -135.4$ kJ mol^{-1}

(b) **Answer:** $\Delta S° = +331.4$ J mol^{-1} K^{-1}

$\Delta S° = \Sigma S_f°(\text{Products}) - \Sigma S_f°(\text{Reactants})$

$\Delta S° = (216.4 + (4 \times 186.8)) - (186.2 + (2 \times 223.0))$

$\Delta S° = +331.4$ J mol^{-1} K^{-1}

(c) **Answer:** $\Delta G_f° = -65.3$ kJ mol^{-1}

$$\Delta G° = \Delta H° - \frac{T\Delta S°}{1000}$$

$\Delta H° = \Delta H_f°(CCl_4) = -135.4$ kJ mol^{-1} (from part (a))

$\Delta S° = \Sigma S_f°(\text{Products}) - \Sigma S_f°(\text{Reactants})$
$\Delta S° = ((216.4) - (5.74 + (2 \times 223.0)))$
$\Delta S° = -235.3$ J mol^{-1} K^{-1}

$$\Delta G° = -135.4 - \frac{((298\text{K}) \times (-235.3 \text{ J mol}^{-1} \text{ K}^{-1}))}{1000}$$

$\Delta G° = -65.3$ kJ mol^{-1}

The $\Delta G°$ of the reaction is the $\Delta G_f°$ of CCl$_4$. Both of the reactants are elements with a $\Delta G_f°$ of zero.

(d) **Answer:** $K = 3 \times 10^{11}$

$\Delta G° = -RT \ln K$

$\ln K = -\dfrac{(65,300 \text{ J mol}^{-1})}{(8.314 \text{ J mol}^{-1} \text{ K}^{-1} \times 298\text{K})} = 26.3$

$K = e^{26.3}$ ('INVERSE' ln 26.3) $= 3 \times 10^{11}$

Chapter 3

Part C: Equations

This part requires that you write 5 reactions equations (chosen from 8) *in net ionic form*. There will always be a reaction. The products will be different from the reactants in every case.

Each reaction is worth 3 points. One (1) point is given for writing the reactant formulas correctly. A maximum of two (2) points is given as (full or partial) credit for correctly predicting the products. Remember: to receive this credit the equation *must* be in net ionic form.

Write out only 5 reactions. Even if you do more, the first 5 will be scored. This part is worth 15 points, and should be given only 10–15 minutes.

The equations should *not* be balanced, and the states are not required. However, balancing the equations and putting in the states may help predict the products, and there is no rule against it.

1. **General Considerations**

    a. Common polyatomic anions and cations.

$C_2H_3O_2^-$	acetate	CO_3^{2-}	carbonate	OH^-	hydroxide
$C_2O_4^{2-}$	oxalate	$Cr_2O_7^{2-}$	dichromate	PO_3^{3-}	phosphite
ClO^-	hypochlorite	CrO_4^{2-}	chromate	PO_4^{3-}	phosphate
ClO_2^-	chlorite	$HCOO^-$	formate	HSO_3^-	bisulfite
ClO_3^-	chlorate	MnO_4^-	permanganate	SO_3^{2-}	sulfite
ClO_4^-	perchlorate	NH_4^+	ammonium	SO_4^{2-}	sulfate
CN^-	cyanide	NO_2^-	nitrite		
HCO_3^-	bicarbonate	NO_3^-	nitrate		

    b. An isolated element which is not an ion has no charge.
       $Na°$ $Mg°$ $Cu°$ $Cl_2°$

    c. There are 7 diatomic elements. The are:

bromine	Br_2	iodine	I_2	nitrogen	N_2	chlorine	Cl_2
hydrogen	H_2	oxygen	O_2	fluorine	F_2		

    (Diatomic R. **BrINClHOF**)

2. **Solubility Rules for Salts in Water Solutions**

   1. **SOLUBLE:** All salts containing:
      ammonium, NH_4^+, and the *Group 1A ions*, Li^+, Na^+, K^+, Rb^+, Cs^+.

   2. **SOLUBLE:** All salts containing:
      nitrate NO_3^-, *acetate*, $C_2H_3O_2^-$, and *perchlorate*, ClO_4^-.

   3. **SOLUBLE:** All salts—see rule 5 for exceptions—containing:
      chloride, Cl^-, *bromide*, Br^-, and *iodide*, I^-.

   4. **SOLUBLE:** All salts containing *sulfates*, SO_4^{2-}
      Exceptions include barium sulfate, $BaSO_4$, calcium sulfate, $CaSO_4$, lead(II) sulfate, $PbSO_4$, and strontium sulfate, $SrSO_4$.

   5. **INSOLUBLE:** All salts containing:
      silver ions, Ag^+, *lead ions*, Pb^{2+}, and *mercury(I) ions*, Hg_2^{2+}.
      Examples are $AgBr$, PbI_2 and Hg_2Cl_2.

   6. **INSOLUBLE:** All salts containing:
      carbonates, CO_3^{2-}, *chromates*, CrO_4^{2-}, *hydroxides*, OH^-, *oxides*, O^{2-}, *phosphates*, PO_4^{3-}, and *sulfides*, S^{2-}.
      However:
      Group 2A *chromates*, except for $BaCrO_4$, are **soluble**.
      Group 2A *hydroxides*, except for $Mg(OH)_2$, are **soluble**.

   Good short cut:
   - With few exceptions, the solubility rules permit assuming: A salt containing an anion with a 1^- charge is **SOLUBLE**. These salts will dissociate into ions. Again, see rule 5 for exceptions.
   - Except for certain sulfates (see rule 4), a salt containing an anion with a 2^- or 3^- charge is **INSOLUBLE**. These salts will **precipitate**.

In a reaction equation when a salt is:
- **SOLUBLE**, the ions are shown *separated*.
    $NaCl(aq) \Rightarrow Na^+ + Cl^-$
- **INSOLUBLE**, the ions are shown *together* as a single formula.
    $PbSO_4(s) \Rightarrow PbSO_4$

In a reaction equation when an acid or base is:
- **STRONG**, the ions are shown *separated*.
    $HCl(aq) \Rightarrow H^+ + Cl^-$
- **WEAK**, the ions are shown *together* as a single formula.
    $H_2SO_3(aq) \Rightarrow H_2SO_3$

Substances that are unchanged as reactants and products in the complete equation—the 'spectator ions'—are not represented in the net ionic equation.

3. **Use of the Solubility Rules—Precipitation Reactions**

   Examples

   1. Aqueous solutions of silver nitrate and sodium iodide are mixed.
      Silver nitrate, $AgNO_3$ (rule 2) is soluble.
      Sodium iodide, NaI (rule 1) is soluble.
      Silver iodide, AgI, is insoluble (rule 3 ahead of rule 4).
      Sodium nitrate, $NaNO_3$, is soluble (rules 1 and 2).
      Complete equation: $Ag^+ + \cancel{NO_3^-} + \cancel{Na^+} + I^- \Rightarrow AgI + \cancel{NO_3^-} + \cancel{Na^+}$
      Net ionic equation: $Ag^+ + I^- \Rightarrow AgI$

   2. Solid aluminum chloride is added to an aqueous solution of potassium chromate.
      Aluminum chloride is soluble (rule 4), but a solid was used in this reaction.
      Potassium chromate and potassium chloride are soluble (rule 1).
      Aluminum chromate is insoluble (rule 5).
      Complete equation: $AlCl_3 + \cancel{K^+} + CrO_4^{2-} \Rightarrow Al_2(CrO_4)_3 + \cancel{K^+} + Cl^-$ (unbalanced)
      Net ionic equation: $AlCl_3 + CrO_4^{2-} \Rightarrow Al_2(CrO_4)_3 + Cl^-$ (unbalanced)

   3. Solutions of barium hydroxide and sulfuric acid are reacted.
      Barium hydroxide is soluble (rule 6–Group 2A hydroxide).
      Sulfuric acid is a strong acid.
      Barium sulfate is insoluble (rule 4 exception).
      Water is not a strong acid.
      Complete equation: $Ba^{2+} + OH^- + H^+ + SO_4^{2-} \Rightarrow BaSO_4 + HOH$
      Net ionic equation: $Ba^{2+} + OH^- + H^+ + SO_4^{2-} \Rightarrow BaSO_4 + HOH$

4. Reactions Involving No Changes in Oxidation States

Replacement Reactions

The products are predicted by exchanging the positive ions of the two reactants. Acids (donating H^+) reacting with bases (accepting the H^+) are representative replacement reactions.

Examples

1. Water solutions of hydrochloric acid and sodium hydroxide are mixed.
$$HCl + NaOH \Rightarrow NaCl + HOH$$
   HCl is a *strong* acid, NaOH is a strong base, and NaCl is a soluble salt; these should be written as ions.

   Complete equation: $H^+ + \cancel{Cl^-} + \cancel{Na^+} + OH^- \Rightarrow \cancel{Na^+} + \cancel{Cl^-} + HOH$
   Net ionic equation: $H^+ + OH^- = HOH$

2. Water solutions of acetic acid and potassium hydroxide are mixed.
$$CH_3COOH + KOH \Rightarrow CH_3COOK + HOH$$
   CH_3COOH is a weak acid and should not be shown separated into ions. KOH is a strong base, and CH_3COOK is a soluble salt. These should be written as separated ions.

   Complete equation: $CH_3COOH + \cancel{K^+} + OH^- \Rightarrow \cancel{K^+} + CH_3COO^- + HOH$
   Net ionic equation: $CH_3COOH + OH^- \Rightarrow CH_3COO^- + HOH$

Metathesis Reactions (Acid and Base Anhydrides)

The products can be predicted after noting that one or both of the reactants is an oxide. The oxides of metals are basic anhydrides and dissolve in water to produce a base. The oxides of nonmetals are acid anhydrides and dissolve in water to produce an acid. An oxide of a metal will react with an oxide of a nonmetal to produce a salt.

Examples

1. Solid calcium oxide is added to water.
   Metal oxide + water $\Rightarrow$ a base
   $CaO(s) + HOH \Rightarrow Ca(OH)_2$ (a soluble hydroxide) $\Rightarrow Ca^{2+} + 2OH^-$
   $CaO(s) + HOH \Rightarrow Ca^{2+} + 2OH^-$

2. Sulfur dioxide is bubbled through water.
   Nonmetal oxide + water $\Rightarrow$ an acid
   $SO_2 + HOH \Rightarrow H_2SO_3$ (a weak acid)

3. Solid calcium oxide is heated in a vessel containing sulfur dioxide.
   Metal oxide + nonmetal oxide $\Rightarrow$ salt
   $CaO + SO_2 \Rightarrow CaSO_3$

Decomposition Reactions

These reactions are the reverse of metathesis reactions and occur when acids, bases or salt are heated.

Examples

1. Solid calcium hydroxide and magnesium hydroxide are heated in an evaporating dish.
    Base $\Rightarrow$ metal oxide + water
    $Ca(OH)_2 \Rightarrow CaO + HOH$
    $Mg(OH)_2 \Rightarrow MgO + H_2O$

2. A solution of carbonic acid is warmed.
    Acid containing oxygen $\Rightarrow$ nonmetal oxide + water
    $H_2CO_3 \Rightarrow CO_2 + HOH$

3. Limestone is heated in a retort.
    Salt containing oxygen $\Rightarrow$ metal oxide + nonmetal oxide
    $CaCO_3 \Rightarrow CaO(s) + CO_2$

Hydrolysis Reactions

Hydrolysis occurs as water reacts with a salt. The water should be written as HOH to allow the ues of an analysis similar to that used for a replacement reaction. Combining the 'H$^+$' of the water with the negative ion of the salt will give the formula of one of the products.

Example

Sodium acetate is dissolved in water.
   $NaC_2H_3O_2 + HOH \Rightarrow HC_2H_3O_2 + NaOH$

Sodium salts are soluble in water (rule 1).
$HC_2H_3O_2$ is not a strong acid and is not shown ionized.
Complete reaction: ~~Na$^+$~~ + $C_2H_3O_2^- + HOH \Rightarrow HC_2H_3O_2 +$ ~~Na$^+$~~ $+ OH^-$
Net ionic reaction: $C_2H_3O_2^- + HOH \Rightarrow HC_2H_3O_2 + OH^-$

Reactions of Coordination Compounds and Ions

Ligands are generally electron-pair donors (Lewis bases) which bond to a central atom that is usually the positive ion of a transition metal. Complex ions and coordination compounds are formed. Important ligands are ammonia, NH_3, cyanide ion, CN^-, and hydroxide ion, OH^-.

Ammonia	'Excess' hydroxide ion	Cyanide ion
$Ag(NH_3)_2^+$	$Al(OH)_4^-$	$Ag(CN)_2^-$
$Cu(NH_3)_4^{2+}$	$Zn(OH)_4^{2-}$	$Cd(CN)_4^{2-}$
$Ni(NH_3)_6^{2+}$	$Cr(OH)_6^{3-}$	$Fe(CN)_6^{3+}$

A useful approximation for the AP Examination is that the number of ligands attached to a central metal ion is often twice the oxidation number of the central metal. It is not important in the scoring of the AP Examination that you know the correct number of ligands. *It is important* that the charge on the ion is correct.

$$AgCl_2^- \qquad Hg(CN)_4^{2-} \qquad Co(NH_3)_6^{3+}$$

When the AP examination is scored there is *no* deduction if the number of ligands is incorrectly written, but there will be a deduction for an incorrectly stated overall charge on the complex ion.

The breakup of complex ions is frequently achieved by adding an acid. The products are the metal ion and the species formed when hydrogen ions from the acid react with the ligand (a Lewis base).

Example

Tetraammine copper II ions are reacted with nitric acid.
$$Cu(NH_3)_4^{2+} + H^+ \Rightarrow Cu^{2+} + NH_4^+$$

Nonaqueous Definitions of Acids and Bases

a. Brönsted reactions involve the transfer of a proton.
b. Lewis reactions involve the formation of a coordinate covalent bond.

Example of a Lewis reaction:
The gases boron trifluoride and ammonia are mixed.
BF_3 has six bonding electrons and is a Lewis *acid* (electron-pair *acceptor*).
NH_3 has a lone electron–pair and is a Lewis *base* (electron-pair *donor*).

$$BF_3 + NH_3 \Rightarrow BF_3NH_3$$

5. **Redox Reactions—Oxidation State Changes by the Oxidizing and the Reducing Agents**

IMPORTANT OXIDIZING AGENTS (Oxidation Number *Decreases*)		IMPORTANT REDUCING AGENTS (Oxidation Number *Increases*)	
Oxidizing Agent (Species Reduced)	**Formed** by the reaction:	**Reducing Agent** (Species Oxidized)	**Formed** by the reaction:
Permanganate, MnO_4^-, in acid	Mn^{2+}	Free halogens (e.g. Br_2)	Hypohalite ions in *dilute* basic sol'n. (e.g. BrO^-)
Permanganate, MnO_4^-, in neutral *or* basic solution	MnO_2	Free halogens (e.g. Cl_2)	Halite ions in *conc.* basic sol'n (e.g. ClO_2^-)
Manganese dioxide, MnO_2, in acid	Mn^{2+}		
HNO_3, concentrated	NO_2	Nitrite ions NO_2^-	Nitrate ions NO_3^-
HNO_3, dilute	NO		
H_2SO_4, hot, concentrated	SO_2	Sulfite ions, SO_3^{2-} or $SO_2(aq)$	Sulfate ions SO_4^{2-}
Metal–ic ions (e.g. Sn^{4+})	Metal–ous ions (e.g. Sn^{2+})	Metal–ous ions (e.g. Fe^{2+})	Metal–ic ions (e.g. Fe^{3+})
Free halogens (e.g. Cl_2)	Halide ions (e.g. Cl^-)	Halide ions (e.g. Br^-)	Free halogen (e.g. Br_2)
Peroxides Na_2O_2, H_2O_2	NaOH, HOH	Free metals (e.g. Na)	Metal ions (e.g. Na^+)
Perchloric acid $HClO_4$	Chloride ion Cl^-		
Dichromate, $Cr_2O_7^{2-}$, in acid	Chromium(III) ion, Cr^{3+}		

Redox Reactions Between an Oxidizing Agent and a Reducing Agent

Redox reactions are generally recognized by:
...familiarization with important reducing agents and oxidizing agents.
...the clue that there is 'added acid' or the solution is 'acidified'.
...the use of the reduction potential reference table supplied with the examination.

Examples

1. Manganese dioxide is added to concentrated hydrochloric acid and heated.
$$MnO_2 + H^+ + Cl^- \Rightarrow Mn^{2+} + Cl_2 + H_2O$$

2. A solution of iron(II) nitrate is added to an acidified solution of potassium permanganate.
$$Fe^{2+} + H^+ + MnO_4^- \Rightarrow Fe^{3+} + Mn^{2+} + H_2O$$

3. Magnesium metal is added to dilute nitric acid. One of the products contains nitrogen with an oxidation number of –3.
$$Mg + H^+ + NO_3^- \Rightarrow Mg^{2+} + NH_3 + H_2O$$

Redox Displacement Reactions

A more reactive element (often in the free state) can displace a less reactive element with similar properties from a compound.

Examples

1. Zinc metal reacts with tin (II) sulfate.
$$Zn + Sn^{2+} \Rightarrow Zn^{2+} + Sn$$

2. Free chlorine reacts with sodium bromide.
$$Cl_2 + Br^- \Rightarrow Cl^- + Br_2$$

3. Solid barium peroxide is added to cold sulfuric acid.
$$BaO_2 + H^+ + SO_4^{2-} \Rightarrow BaSO_4 + H_2O_2$$

Redox Combination Reactions

An oxidizing agent will react with a reducing agent of the *same* element to produce the element at an intermediate oxidation state.

Examples

1. Solutions of potassium iodide, potassium iodate, and dilute sulfuric acid are mixed.

    $I^- + IO_3^- + H^+ \Rightarrow I_2 + H_2O$

2. A piece of iron is added to a solution of iron (III) sulfate.

    $Fe + Fe^{3+} \Rightarrow Fe^{2+}$

Redox Decomposition Reactions

Examples

1. A solution of hydrogen peroxide is catalytically decomposed.

    $H_2O_2 \Rightarrow H_2O + O_2$

2. Chlorates decompose in the presence of heat.

    $KClO_3 \Rightarrow KCl + O_2$

3. Electrolysis decomposes compounds into their elements.

    $H_2O \Rightarrow H_2 + O_2$

6. **Practice Equation Writing**

Write the formula equations for each reaction in net ionic format. Do not balance. There is a reaction in each case. Also, attempt to predict the reaction type that will lead to the correct prediction of the products.

1. Sodium sulfite crystals are added to water.

2. A potassium dichromate solution is added to an aqueous solution of lead(II) nitrate.

3. Calcium carbonate chips are added to excess nitric acid.

4. Hydrogen sulfide gas is bubbled through a solution of cadmium nitrate.

5. Solutions of acetic acid and sodium bicarbonate are mixed.

6. Hydrogen chloride gas is bubbled through water.

7. Solutions of carbon dioxide gas and ammonia gas are mixed.

8. Solid lithium oxide is added to water.

9. Sulfur trioxide gas is bubbled through a solution of sodium hydroxide.

10. A mixture of the solid calcium oxide and solid tetraphosphorous decaoxide is heated.

11. Solid magnesium carbonate is heated in a crucible.

12. Solid phosphorus trichloride is added to water.

13. Solid sodium cyanide is added to water.

14. Solid zinc nitrate is treated with excess sodium hydroxide solution.

15. An aqueous solution of diamminesilver chloride is treated with dilute nitric acid.

16. A suspension of copper(II) hydroxide in water is treated with an excess of ammonia.

17. Gaseous boron hydride gas is mixed with ammonia gas.

18. Boron trifluoride is added to gaseous trimethylamine.

19. Gaseous silane (SiH_4) is burned in excess oxygen.

20. Magnesium ribbon is burned in pure nitrogen gas.

21. An aqueous hydrogen peroxide solution is added to an acidified solution of potassium iodide.

22. Dilute potassium permanganate solution is added to an oxalic acid solution which is acidified with a few drops of sulfuric acid.

23. A solution containing tin(II) ion is added to acidified potassium dichromate.

24. A segment of copper wire is added to dilute nitric acid.

25. Sodium dichromate is added to an acidified solution of sodium iodide.

26. Sodium hydride crystals are added to water.

27. Powdered iron is added to a solution of iron(III) sulfate.

28. Chlorine gas is bubbled through a solution of potassium bromide.

29. Solid sodium is added to water.

30. A dilute solution of sulfuric acid is electrolyzed between platinum electrodes.

Part C: Equations

Sample Answers and Explanations

1. *Notes*: Sodium salts are soluble in water (rule 1).
   Bases react with water for form OH^-.
   EQUATION: $Na_2SO_3 \Rightarrow Na^+ + SO_3^{2-} + HOH \Rightarrow Na^+ + HSO_3^- + OH^-$
   NET IONIC: $Na_2SO_3 \Rightarrow Na^+ + HSO_3^- + OH^-$
   REACTION TYPE: Solubility and hydrolysis

2. *Notes*: Potassium (rule 1) and nitrate (rule 2) salts dissolve completely. The K^+ and NO_3^- ions are 'spectators'.
   EQUATION: $\cancel{K^+} + Cr_2O_7^{2-} + Pb^{2+} + \cancel{NO_3^-} \Rightarrow PbCr_2O_7 + \cancel{K^+} + \cancel{NO_3^-}$
   NET IONIC: $Cr_2O_7^{2-} + Pb^{2+} \Rightarrow PbCr_2O_7$
   REACTION TYPE: Replacement

3. *Notes*: HNO_3 is a strong acid and completely ionizes.
   Carbonic acid, H_2CO_3, decomposes in water to give CO_2 gas bubbles.
   EQUATION: $CaCO_3(s) + H^+ + \cancel{NO_3^-} \Rightarrow Ca^{2+} + \cancel{NO_3^-} + H_2CO_3$
   NET IONIC: $CaCO_3(s) + H^+ \Rightarrow Ca^{2+} + CO_2 + H_2O$
   REACTION TYPE: Replacement

4. *Notes*: Nitrate salts are soluble (rule 2).
   Many sulfides are insoluble (rule 6).
   EQUATION: $Cd^{2+} + \cancel{NO_3^-} + H_2S \Rightarrow CdS + H^+ + \cancel{NO_3^-}$
   NET IONIC: $Cd^{2+} + H_2S \Rightarrow CdS + H^+$
   REACTION TYPE: Replacement

5. *Notes*: Acid-base replacement (Brönsted proton transfer).
   H_2CO_3 decomposes to form CO_2 and H_2O.
   EQUATION: $HC_2H_3O_2 + \cancel{Na^+} + HCO_3^- \Rightarrow C_2H_3O_2^- + \cancel{Na^+} + H_2CO_3$
   NET IONIC: $HC_2H_3O_2 + HCO_3^- \Rightarrow C_2H_3O_2^- + CO_2 + H_2O$
   REACTION TYPE: Replacement

6. *Notes*: Acid-base replacement (Brönsted proton transfer).
   NET IONIC: $HCl + HOH \Rightarrow H_3O^+ + Cl^-$
   REACTION TYPE: Replacement

7. *Notes*: Acid-base replacement (Brönsted proton transfer).
   H_2CO_3, carbonic acid, is $CO_2 + H_2O$.
   EQUATIONS: 1. $CO_2 + H_2O \Leftrightarrow H_2CO_3$
   2. $NH_3(aq) + H_2CO_3(aq) \Rightarrow NH_4^+ + HCO_3^-$
   NET IONIC: $CO_2 + H_2O + NH_3(aq) \Rightarrow NH_4^+ + HCO_3^-$
   REACTION TYPE: Replacement

8. **Notes:** Li$_2$O is a basic anhydride.
   LiOH is a strong base and completely dissociates.
   EQUATION: Li$_2$O + HOH $\Rightarrow$ LiOH(aq) $\Rightarrow$ Li$^+$ + OH$^-$
   NET IONIC: Li$_2$O + HOH $\Rightarrow$ Li$^+$ + OH$^-$
   REACTION TYPE: Metathesis

9. **Notes:** SO$_3$ is an acid anhydride.
   H$_2$SO$_4$ is a strong acid and sodium hydroxide is a strong base.
   EQUATION:
   1. SO$_3$(g) + H$_2$O $\Rightarrow$ H$_2$SO$_4$ $\Rightarrow$ H$^+$ + HSO$_4^-$
   2. H$^+$ + ~~Na$^+$~~ + OH$^-$ $\Rightarrow$ HOH + ~~Na$^+$~~

   NET IONIC: SO$_3$ + OH$^-$ $\Rightarrow$ HSO$_4^-$
   REACTION TYPE: Metathesis
   1. Nonmetal oxide in water gives an acid.
   2. Acid-base (Brönsted proton transfer).

10. **Notes:** Metal oxide + nonmetal oxide = salt
    There is *no* change in the oxidation number.
    EQUATION: CaO(s) + P$_4$O$_{10}$(s) $\Rightarrow$ Ca$_3$(PO$_4$)$_2$(s)
    REACTION TYPE: Metathesis

11. **Notes:** Heating a metal carbonate yields a metal oxide and CO$_2$.
    EQUATION: MgCO$_3$ $\Rightarrow$ MgO + CO$_2$
    REACTION TYPE: Decomposition

12. **Notes:** HCl is a strong acid and completely ionizes.
    EQUATION: PCl$_3$ + HOH $\Rightarrow$ H$^+$ + Cl$^-$ + P(OH)$_3$
    or: PCl$_3$ + HOH $\Rightarrow$ H$^+$ + Cl$^-$ + H$_3$PO$_3$
    REACTION TYPE: Hydrolysis or replacement

13. **Notes:** HCN is not a strong acid.
    EQUATION: NaCN + HOH $\Rightarrow$ HCN + Na$^+$ + OH$^-$
    REACTION TYPE: Hydrolysis of a base

14. **Notes:** Key words are 'excess sodium hydroxide'.
    Sodium and nitrate salts are soluble.
    EQUATION: Zn(NO$_3$)$_2$ + ~~Na$^+$~~ + OH$^-$ $\Rightarrow$ Zn(OH)$_4^{2-}$ + ~~Na$^+$~~ + NO$_3^-$
    NET IONIC: Zn(NO$_3$)$_2$ + OH$^-$ $\Rightarrow$ Zn(OH)$_4^{2-}$ + NO$_3^-$
    REACTION TYPE: Reactions of coordination compounds and ions.

15. *Notes*: The breakup of complex ions is frequently achieved by adding an acid which reacts with the basic ligand.
    Many silver salts are insoluble (rule 3).
    EQUATION: $Ag(NH_3)_2^+ + Cl^- + H^+ + \cancel{NO_3^-} \Rightarrow AgCl + NH_4^+ + \cancel{NO_3^-}$
    NET IONIC: $Ag(NH_3)_2^+ + Cl^- + H^+ \Rightarrow AgCl + NH_4^+$
    REACTION TYPE: Reactions of complex ions

16. *Notes*: The key words are 'excess ammonia'.
    EQUATION: $Cu(OH)_2(s) + NH_3(aq) \Rightarrow Cu(NH_3)_4^{2+} + OH^-$
    REACTION TYPE: Reactions of coordination compounds

17. *Notes*: BH_3 has 6 outer electrons. NH_3 has 8 including a 'lone pair'.
    A coordinate covalent bond is formed.
    EQUATION: $BH_3 + NH_3 \Rightarrow BH_3NH_3$
    REACTION TYPE: Lewis acids and bases

18. *Notes*: BF_3 has 6 outer electrons. $(CH_3)_3N$ has 8 including a 'lone pair'.
    A coordinate covalent bond is formed.
    EQUATION: $BF_3 + (CH_3)_3N \Rightarrow F_3BN(CH_3)_3$
    REACTION TYPE: Lewis acids and bases

19. *Notes*: Combustion produces oxides of the 'fuel'.
    In excess oxygen the highest oxidation state oxide forms.
    EQUATION: $SiH_4 + O_2 \Rightarrow SiO_2 + H_2O$
    REACTION TYPE: Reaction between reducing agent and oxidizing agent

20. *Notes*: Metals react with nonmetals to form salts.
    EQUATION: $Mg(s) + N_2(g) \Rightarrow Mg_3N_2(s)$
    REACTION TYPE: Reaction between reducing agent and oxidizing agent.
    Redox combination

21. *Notes*: Peroxide *in acid* is an important oxidizing agent.
    Halide ions are reducing agents.
    Potassium salts are soluble (rule 1).
    EQUATION: $H_2O_2 + H^+ + \cancel{K^+} + I^- \Rightarrow H_2O + \cancel{K^+} + I_2$
    NET IONIC: $H_2O_2 + H^+ + I^- \Rightarrow H_2O + I_2$
    REACTION TYPE: Reaction between oxidizing agent and reducing agent

22. *Notes*: Permanganate *in acid* is an important oxidizing agent.
    Oxalic acid, $H_2C_2O_4$, is not a strong acid.
    H_2CO_3 decomposes to form CO_2 and H_2O.
    EQUATION: $MnO_4^- + H^+ + H_2C_2O_4 \Rightarrow Mn^{2+} + CO_2 + H_2O$
    REACTION TYPE: Reaction between oxidizing agent and reducing agent

23. *Notes*: Dichromate *in acid* is an important oxidizing agent.
    EQUATION: $Sn^{2+} + H^+ + Cr_2O_7^{2-} \Rightarrow Sn^{4+} + Cr^{3+} + H_2O$
    REACTION TYPE: Reaction between reducing agent and oxidizing agent

24. *Notes*: Nitric acid is a strong acid and an important oxidizing agent.
    EQUATION: $Cu + H^+ + NO_3^- \Rightarrow Cu^{2+} + NO + H_2O$
    REACTION TYPE: Reaction between reducing agent and oxidizing agent

25. *Notes*: Dichromate in *acid* is an important oxidizing agent.
    Sodium salts are soluble in water (rule 1).
    EQUATION: $\cancel{Na^+} + Cr_2O_7^{2-} + H^+ + I^- \Rightarrow \cancel{Na^+} + Cr^{3+} + I_2 + H_2O$
    NET IONIC: $Cr_2O_7^{2-} + H^+ + I^- \Rightarrow Cr^{3+} + I_2 + H_2O$
    REACTION TYPE: Reaction between oxidizing agent and a reducing agent

26. *Notes*: NaOH completely dissociates.
    EQUATION: $NaH(s) + HOH \Rightarrow Na^+ + OH^- + H_2(g)$
    REACTION TYPE: Reaction between oxidizing agent and a reducing agent.
    Metal hydride + HOH $\Rightarrow$ base + hydrogen.

27. *Notes*: Fe^{2+} is the intermediate oxidation state or iron.
    EQUATION: $Fe + Fe^{3+} \Rightarrow Fe^{2+}$
    REACTION TYPE: Redox combination

28. *Notes*: The more reactive element displaces the less reactive element.
    For the halogen reactivity: $F_2 > Cl_2 > Br_2 > I_2$.
    Potassium salts completely dissolve (rule 1).
    EQUATION: $Cl_2 + \cancel{K^+} + Br^- \Rightarrow \cancel{K^+} + Cl^- + Br_2$
    NET IONIC: $Cl_2 + Br^- \Rightarrow Cl^- + Br_2$
    REACTION TYPE: Redox replacement

29. *Notes*: The more reactive element, Na, displaces a less reactive element, H_2 in the compound.
    Strong bases dissociate completely.
    EQUATION: $Na + HOH \Rightarrow Na^+ + OH^- + H_2$
    REACTION TYPE: Redox replacement

30. *Notes*: Electrolysis decomposes compounds into their elements. In this case the voltage required to decompose water is less than the voltage required to decompose H_2SO_4.
    EQUATION: $H_2O \Rightarrow H_2 + O_2$
    REACTION TYPE: Redox decomposition (electrolysis)

Chapter 4

Part D: Essays

The essays require an ability to present ideas in a logical order and making use of defined chemical principles to explain the observations that are cited in the question. Part D consists of five topics—three must be answered and one is a required question—and is worth 45 percent of Section II. Even if you do more than three questions, only the first three will be scored. About 40 minutes should be spent on this part, or about 10–15 minutes for each question.

In order to conserve time, the answers to the essays should be as concise as possible. For these questions:

- Think before you start writing.

- Circle the words that tell you what to do (i.e. compare, contrast, describe, explain).

- Define the chemical terms used in the question. Most essays will be correctly answered by using defined words properly.

- Be organized. Start with a thesis statement, provide supporting points, and finish with a conclusion.

- Be specific. Use only relevant facts in your answer.

- Stick to the point of the question. Points can be lost by giving an answer that is too general.

- Limit yourself to one thought per sentence. The resulting sentences will be of different lengths and will be in a more conversational tone.

- Do not get hung up so long on one question that time is not available for other questions.

- Do not change essay answers if time is running out. When one sentence is changed there is a good chance that other sentences will need to be changed.

This table shows the questions asked in Part D for the past several years, not necessarily in the order in which they were asked. The essays chosen balance with the topics in Parts A and B so that all the major areas are tested in a somewhat balanced manner.

Year	Part D				
1976	*Atom* Periodicity	*Bond* Octets	*Thermo* ΔH, ΔS, ΔG	*States* Ideal Gases	*Kinetics* Rate
1977	*Atom* Affinity	*Bond* Major types	*Bond* Coordinates	*Bond* Isomers	*Equilibrium* LeChatelier
1978	*Atom* Bohr Model	*Atom/Bond* Polarity, I.E.	*Bond* Isomers	*Stoich* Colligative	*Equilibrium* Hydrolysis
1979	*Bond* Shapes	*Bond* Dipoles	*Laboratory* Analysis	*Kinetics* Rates	*Equilibrium* Acid-Base
1980	*Atom* Structure	*Laboratory* Solutions	*Kinetics* Pot. Energy	*Equilibrium* LeChatelier	*Thermo* Entropy
1981	*Atom* Spectra	*Stoich* Empirical	*Equilibrium* Hydrolysis	*Thermo* Free Energy	*Electrochem* Electroysis
1982	*Atom* Periodicity	*Bond* VSEPR	*States* Ideal	*Laboratory* Solutions	*Equilibrium* Titration
1983	*Bond* Coordinates	*Bond* Molec. Orb.	*Equilibrium* Buffers	*Kinetics* Order	*Electrochem* Voltaic cell
1984	*Atom* Metals	*States* Vap. Press.	*Laboratory* Solutions	*States* Gases	*Equilibrium* Titration
1985	*Atom* Periodicity	*Bond* Dipoles	*Laboratory* Solutions	*Kinetics* Rate Law	*Thermo* ΔH, ΔS, ΔG
1986	*Atom* I.E., ΔH	*Bond* Properties	*Bond* Oxyacids	*Kinetics* Rate Law	*Electrochem* Reactions
1987	*Atom* Uncertainty	*Atom* Periodicity	*Bond* Dissociation	*Thermo* ΔH, ΔS, ΔG	*Electrochem* Electrolysis
1988	*Bond* Properties	*States* Phases	*Equilibrium* LeChatelier	*Equilibrium* Titration	*Thermo* Lab ΔH
1989	*Atom* Decay	*Bond* Shapes	*Bond* Melt Prop.	*Laboratory* Metal React	*Kinetics* Reaction Rate

Starting with the 1990 examination, the format of Part D changed. To be assured that all of the topics of the AP Chemistry syllabus are covered by the examination, the rules of the test make the first question of Part D—Question #5 of Section II—*required*.

Year	Part D				
	Required	**Choose 2 of 4 questions**			
1990	*Bond* Length–angle	*Atom* Ioniz. energy	*Equilibrium* Acid-Base	*Kinetics* Rate law	*Laboratory* Gravimetric
1991	*Thermo* ΔH, ΔS, ΔG	*Atom* Nuclear	*Bond* MP and type	*Electrochem* Voltaic cells	*Laboratory* MM of liquid
1992	*Kinetics* Rates	*Bond* Hybrid, angle	*Equilibrium* Buffers	*Laboratory* unknown cmpd	*States* Attract. forces

Sample Essay Questions

1. An empirical rule commonly known as Hund's Rule states that "in a given atom, so long as the Pauli Exclusion Principle permits, electrons in the same subshell will occupy orbitals with different values of *m* and their spins will not pair up."

    (a) What is the Pauli Exclusion Principle?

    (b) Explain how Hund's Rule is applied to the electron structure of many–electron atoms.

    (c) Some atoms are paramagnetic and others are diamagnetic. What is the difference between the electronic structure of elements which are paramagnetic and those that are diamagnetic?

    (d) Hund's Rule has been confirmed by measurements of the magnetic properties of elements. Using selected Period 2 elements as examples, show how the confirmation of Hund's Rule has been accomplished.

2. The electron affinities, in kJ mol^{-1}, are given for five elements in Period 2.

B = -23	C = – 123	N = 0	O = –142	F = – 322

    (a) Why does the electron affinity of the elements generally increase with atomic number in Period 2.

    (b) Why is the electron affinity of nitrogen zero, while carbon's and oxygen's are substantial?

3. In a P_4 molecule, the phosphorus atoms are at the corners of a tetrahedron. Each phosphorus atom is bonded to the three others. All the bonds have the same energy.

    (a) Draw a diagram showing the shape of a P_4 molecule.

    (b) What is the bond angle for each P—P—P bond?

    (c) Draw a diagram showing the shape of a PH_3 molecule.

    (d) Account for the difference between P—P—P bond angle and the H—P—H bond angle.

4. The strength of bonding of ligands in complex ions decreases in the order:
$$CN^- > NO_2^- > NH_3 > H_2O > OH^- > F^- > Cl^- > Br^- > I^-.$$
Thus cyanide ion, CN^-, is at the head of the list and is said to be a very strong ligand. At the other end iodide ion, I^-, is a very weak ligand.

(a) Draw the Lewis electron-dot structures of the cyanide ion and the nitrite ion.

(b) How many sigma, σ, and how many pi, π, bonds are in:

...a cyanide ion?

...a nitrite ion?

(c) What characteristic of the structure of ligands causes them to form complex ions?

(d) Explain where carbon monoxide, CO, would fit into the series of bonding strengths of ligands.

5. Iodate ion, IO_3^-, can be used to titrate tantalum(I) ion, Tl^+, in a concentrated solution of HCl.

$$IO_3^- + Tl^+ + Cl^- \Rightarrow ICl_2^- + Tl^{3+}$$

(a) Write the correctly balanced:

(1) oxidation half-reaction

(2) reduction halfreaction

(3) net ionic reaction

(b) Which is the reducing agent and which is the oxidizing agent in the reaction?

(c) One of the reactants can be made 1.50 M, and one of the products can be made 0.10 M. Which reactant and which product should be chosen to maximize the net cell voltage, $\mathcal{E}°$? Explain your reasoning.

(d) A measurement or a calculation of the net cell voltage, $\mathcal{E}°$, can be used to determine two additional properties of the reaction. List the two properties of the reaction that would be determined.

6. Four beakers containing 0.10 M solutions are prepared.
   AgNO₃   Al(NO₃)₃   LiNO₃   Zn(NO₃)₂

   (a) A 6.0 M solution of sodium hydroxide, NaOH, is added to each solution. Write balanced ionic equations for the solutions where a precipitation reaction occurs.

   (b) Excess NaOH is added to the beakers where a precipitate has formed. Write balanced net ionic equations for the precipitate(s) that undergo further reaction.

7. The equilibrium shown by the equation at 25°C and a constant volume of 1.0 liter.

   $$C_{graphite}(s) + 2H_2(g) \Leftrightarrow CH_4(g) \quad \Delta H° = -75 \text{ kJ}$$

   State whether the molarity, M, of *each* of the three constituents of the system shows a net increase, a net decrease or remains constant as the system comes back to equilibrium.

   (a) Increase the temperature.
   (b) Remove some hydrogen, H_2.
   (c) Add some graphite, C.
   (d) Decrease the pressure.
   (e) Add some methane, CH_4.

8. The weight percent of NaHCO₃ in a mixture of sodium hydrogen carbonate, NaHCO₃, and sodium chloride, NaCl, was determined by heating the mixture. Upon heating, the mixture lost weight equal to the weight of the carbon dioxide and water released by the NaHCO₃.

   (a) Write a balanced reaction equation.

   (b) If time was not taken to bring the residue to constant weight by heating the sample long enough, would the reported percent sodium bicarbonate be too high or too low? Explain your conclusion.

   (c) Before the sample can be weighed, it must be cooled to about room temperature. Would the reported percent sodium bicarbonate be too high or too low if the cooling process were allowed to occur in the room overnight? Explain your conclusion.

9. Phosgene, $COCl_2$, is produced in the gas phase from carbon monoxide, CO, and chlorine, Cl_2.

$$CO(g) + Cl_2(g) \Rightarrow COCl_2(g)$$

The reaction proceeds by the mechanism:

1.	Cl_2	$\Leftrightarrow$ 2Cl	(fast equilibrium)
2.	Cl + CO	$\Leftrightarrow$ COCl	(fast equilibrium)
3.	$COCl + Cl_2$	$\Rightarrow$ $COCl_2$ + Cl	(slow)
4.	2Cl	$\Leftrightarrow$ Cl_2	(fast equilibrium)

(a) Write the rate law equation in terms of substances that appear in the overall reaction.

(b) What are the units of the rate law constant, k?

(c) An increase of 10°C will possibly double the rate of a chemical reaction. Explain what occurs on the molecular level that accounts for this rate change.

(d) The value of the rate constant, k, can be graphed at different Kelvin temperatures, T, using data obtained in the laboratory. From this data the activation energy, E_a, can be determined.

    1. What should be plotted as the independent variable (x-axis) and what should be plotted as the dependent variable (y-axis)?

    2. How would the graph be used to determine the activation energy for the reaction?

10. For the reaction:

$$Fe_2O_3(s) + 3C(s) \Leftrightarrow 2Fe(s) + 3CO(g)$$

$\Delta H°$, $\Delta S°$, and $\Delta G°$ are all positive when the substances are in their standard states at 25°C.

(a) What is the physical significance of the signs of $\Delta H°$, $\Delta S°$, and $\Delta G°$ for this reaction?

(b) Which of the substances would exist in the highest amounts in an equilibrium mixture at 25°C? Explain how you arrived at your answer.

(c) This reaction is used as a step in the recovery of iron from its ore. Use thermodynamic concepts to explain how the yield of iron can be maximized.

11. Deviations from ideal behavior for nitrogen, N_2, hydrogen, H_2, and carbon dioxide, CO_2, are shown in the graph. Ideal gases have a PV/nRT ratio equal to 1.0.

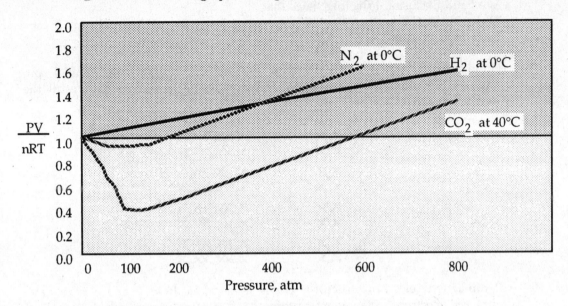

(a) There are two ways that real gases deviate from ideal behavior. List them.

(b) In the temperature range shown (0–40°C), compare the molecular attractive forces of nitrogen, hydrogen, and carbon dioxide. How does the data plotted in the graph justify the comparison?

(c) At high pressures, the PV/nRT ratio of all three gases exceeds 1.0. Explain this in terms of the kinetic molecular theory.

(d) At 600 atm, the PV/nRT ratio of nitrogen is greater than that of hydrogen. Does this mean that the attraction between hydrogen molecules is greater than that of hydrogen? Justify your conclusion.

Sample Answers and Explanations

1. (a) The Pauli Exclusion Principle states that:
   "Two, but not more than two electrons can occupy an orbital, and only if they have opposite spin."

   (b) Hund's Rule, restated, is that electrons will half-fill orbitals with parallel spin until all the orbitals are half-filled. Subsequent electrons will pair until the subshell is filled.

Carbon	2s ⊗	2p	⊗ ○ ○	
Nitrogen	2s ⊗	2p	⊗ ⊗ ⊗ (half-filled)	
Oxygen	2s ⊗	2p	⊗ ⊗ ○	
Neon	2s ⊗	2p	⊗ ⊗ ⊗ ⊗	

   (c) Paramagnetic electron structures create a magnetic field because they have unpaired electrons. Diamagnetic atoms have electron structures where all of the electrons are paired. Electron pairs cancel the magnetic field.

   (d) The magnetic fields of carbon and nitrogen show paramagnetism because of 2 (carbon) and 3 (nitrogen) unpaired electrons. Neon, with all electrons paired, would be diamagnetic.

2. (a) Electron affinity is a measurement of the amount of energy released when an electron is added to a neutral, isolated gaseous atom.
   The electron affinity increases with increasing atomic number in a period because electrons are being added to the valence shell at the same time that the number of protons attracting them is increasing.

   (b) The electron affinity of nitrogen is zero because nitrogen has a stable electron structure. The $2s$-subshell is filled and the $2p$-subshell is exactly half-filled. Filled and half-filled electron structures are especially stable.

3. (a)

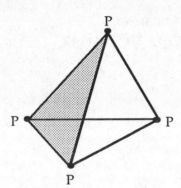

(b) Each face of the tetrahedron an equilateral triangle, so the P—P—P bond angle is 60°.

(c)

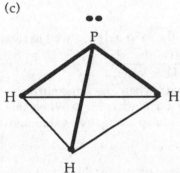

(d) The bond angle is < 109.5°. The H—P—H bond angle is inside a tetrahedron. The bond angle for this location is usually 109.5°, but the lone electron pair will reduce the angle.

4. (a)

cyanide ion $[:C\equiv N:]^-$

nitrite ion $[\overset{..}{\underset{..}{:O}}-N=\overset{..}{\underset{..}{O}}:]^-$

(b) The cyanide ion has one σ bond and two π bonds.
The nitrite ion has two σ bonds and one π bond.

(c) Ligands have lone electron pairs which participate in coordinate covalent bonding with d-orbitals of a metal ion to form complex ions.

(d) Carbon monoxide is structurally similar to the cyanide ion, CN^-, and would fit into the series of bonding strength of ligands near CN^-.

carbon monoxide $:C\equiv O:$

5. (a)
Oxidation: $Tl^+ \Rightarrow Tl^{3+} + 2e^-$
Reduction: $IO_3^- + 2Cl^- + 6H^+ + 4e^- \Rightarrow ICl_2^- + 3H_2O$
Net ionic: $IO_3^- + 2Tl^+ + 2Cl^- + 6H^+ \Leftrightarrow ICl_2^- + 2Tl^{3+} + 3H_2O$

(b) The reducing agent, Tl^+, is oxidized.
The oxidizing agent, IO_3^-, is reduced.

(c) The Nernst equation for this reaction is:

$$\varepsilon = \varepsilon^\circ - \frac{0.059}{4} \log \frac{[ICl_2^-][Tl^{3+}]^2}{[IO_3^-][Tl^+]^2[Cl^-]^2[H^+]^6}$$

This redox reaction is not at equilibrium until the ratio of [Products] to [Reactants] is equal to the equilibrium constant, K. $\left(K = \frac{[Products]}{[Reactants]} \right)$

High values of the net cell potential, ε, occur when the concentration of the products is low and the concentration of the reactants is high. When this occurs the log expression in the Nernst equation will have a negative value and will be added (a negative times a negative) to ε°

The $[Tl^{3+}]^2$ and $[H^+]^6$ have the highest exponents in the equilibrium law expression, and these should be 0.10 M and 1.50 M, respectively.

(d) The standard free energy change, ΔG°, can be calculated from the net cell potential using the equation:

$$\Delta G^\circ = -nF\varepsilon^\circ$$

The equilibrium constant, K can be calculated from the standard net cell potential using the equation:

$$\log K = \frac{n}{0.059} \varepsilon^\circ$$

6. (a) $Ag^+(aq) + OH^-(aq) \Rightarrow AgOH(s)$
$Al^{3+}(aq) + 3OH^-(aq) \Rightarrow Al(OH)_3(s)$
$Zn^{2+}(aq) + 2OH^-(aq) \Rightarrow Zn(OH)_2(s)$

(b) Aluminum and zinc hydroxides are amphoteric.
$Al(OH)_3(s) + OH^-(aq) \Rightarrow [Al(OH)_4]^-(aq)$
$Zn(OH)_2(s) + 2OH^-(aq) \Rightarrow [Zn(OH)_4]^{2-}(aq)$

7.

	[C(s)], M	[H₂(g)], M	[CH₄(g)], M
(a)	Remains the same	Increases	Decreases
(b)	Remains the same	Net decrease. (Decreases, then increases)	Decreases
(c)	Remains the same	Remains the same	Remains the same
(d)	Remains the same	Increases	Decreases
(e)	Remains the same	Increases	Net increase. (Increases, then decreases)

Notes:
- (a) An increase in temperature favors the endothermic reaction; in this case the reverse reaction. The concentration of a solid is constant.
- (b) The removal of H₂ slows down the forward reaction rate.
- (c) Addition of a solid does not affect the equilibrium.
- (d) Decreasing the pressure favors the reaction that will increase the number of moles of gas.
- (e) The addition of CH₄ speeds up the forward reaction rate.

8. (a) $2NaHCO_3(s) \Leftrightarrow Na_2CO_3(s) + CO_2(g) + H_2O(g)$

(b) The sample, by not being heated enough, will not evolve CO_2 and H_2O completely. If the amount of gases evolved is too low the computed (from the loss in weight of the sodium bicarbonate) percent $NaHCO_3$ would also be low.

(c) During the night some CO_2 and H_2O from the air would be absorbed by the residue, Na_2CO_3, and its mass would increase. The computed percent $NaHCO_3$ (from the loss in weight) would be too low. The sample should be reheated and cooled just before weighing.

9. (a) Rate = $k_3[COCl][Cl_2]$ (from slow step)

$[COCl] = K_2[Cl][CO]$ (from reaction 2)

$[Cl] = \sqrt{K_1}\sqrt{[Cl_2]}$ (from reaction 1)

Rate = $K_2 k_3 \sqrt{K_1}[CO][Cl_2]^{3/2} = k[CO][Cl_2]^{3/2}$

(b) The units of k must allow the rate units to be mol L⁻¹ time⁻¹.
For this reaction, the units of k are $L^{3/2}$ mol$^{-3/2}$ time⁻¹

(c) At the molecular level, when the temperature is increased the average kinetic energy of the molecules increases. This results in more collisions between molecules because their molecular velocity is higher. But, more importantly, there is an increase in the number of molecules with sufficient activation energy to react when a collision occurs.

(d) A form of the Arrhenius equation is:

$$\ln \frac{k_2}{k_1} = -\frac{E_a}{R}\left[\frac{1}{T_2} - \frac{1}{T_1}\right]$$

1. $\ln k$ should be the dependent variable (y-axis).
   $1/T$ should be the independent variable (x-axis).

2. The slope of the line is $-\frac{E_a}{R}$.

10. (a) When $\Delta H°$ is positive, the reaction is endothermic.
    When $\Delta S°$ is positive, the products are more chaotic.
    When $\Delta G°$ is positive, the reaction is nonspontaneous.

    (b) The substances in the highest amounts are the reactants, Fe_2O_3 and C. The reaction is not spontaneous and little, if any, product will be formed.

    (c) $\Delta G° = \Delta H° - T\Delta S°$
    Increase the temperature of the process. The positive entropy change is favorable for spontaneous reaction, and the $T\Delta S°$ term will be higher at higher temperatures. When $T\Delta S°$ is greater than $\Delta H°$ the free energy change, $\Delta G°$, will be negative.

11. (a) 1. Real gas molecules have volume.
           Ideal gas molecules take up no space.

        2. Real gas molecules attract each other.
           Ideal gases have no attractions between molecules.

    (b) Carbon dioxide molecules have a higher van der Waals attraction than H_2 or N_2. The molecules deflect each other and their paths to the wall are curved rather than straight. As a result they hit the wall less frequently, and the pressure is lower at each temperature.

    (c) At a higher pressure, the molecular volume becomes more significant. The molecules are closer together and the effective volume (void space) is smaller than expected. There will be a higher collision rate with the walls of the container than with an ideal gas.

    (d) The attraction between hydrogen molecules is not greater than that of nitrogen. Hydrogen molecules have fewer electrons, so the molecular attraction between H_2 molecules is *lower* than those of N_2.
    The reason the PV/nRT ratio for nitrogen is higher than hydrogen is that N_2 is a larger molecule.

Appendix

A. Summary of Formulas

Chapter 1: Atomic Structure and Periodicity

HALF LIFE (A FIRST ORDER KINETICS REACTION)

$$\ln \frac{N_t}{N_o} = -\lambda t \qquad \lambda = \frac{0.693}{t_{1/2}} \qquad \ln \frac{N_t}{N_o} = -\frac{0.693 t}{t_{1/2}}$$

N_t is the number of moles of the isotope at time t.
N_o is the original number of moles of the isotope.
$\frac{N_t}{N_o}$ is the mole fraction of the isotope remaining at time t.
λ is the decay constant and $t_{1/2}$ is the half-life.

Chapter 3: Stoichiometry

MOLE LINK

The formula H₂O could be read to mean that there are:

...half as many moles of oxygen as moles of hydrogen.   ($n_{oxygen} = \frac{1}{2} n_{hydrogen}$)

...twice as many moles of hydrogen as moles of oxygen.   ($n_{hydrogen} = 2\, n_{oxygen}$)

...three times as many moles of atoms as moles of molecules.   ($n_{atoms} = 3\, n_{molecules}$)

SOLUTION CONCENTRATION

$$\text{Molarity (M)} = \frac{\text{moles solute}}{\text{liter solution}} = \frac{(n)}{(V)}$$

$$\text{Molality (m)} = \frac{\text{moles solute}}{\text{kilogram solvent}} = \frac{(n)}{(kg)}$$

$$\text{Mole fraction (x)} = \frac{\text{moles solute}}{\text{moles solution}}$$

DILUTION

$$M_{final} = \frac{V_{initial} \times M_{initial}}{V_{final}}$$

COLLIGATIVE PROPERTIES

fp and bp change $\Delta T_f = K_f m$ $\Delta T_b = K_b m$

Raoult's Law $P_{solvent} = x_{solvent} P^\circ$ $\Delta P_{solvent} = x_{solute} P^\circ$

Appendix

Chapter 4: States of Matter

GRAHAM'S LAW

$$\frac{\text{Rate of effusion of Gas A}}{\text{Rate of effusion of Gas B}} = \sqrt{\frac{\text{Molar Mass of Gas B}}{\text{Molar Mass of Gas A}}}$$

IDEAL GAS LAW; DENSITY OF A GAS; MOLECULAR WEIGHT OF A GAS

$$PV = [n]RT$$

$$D = \frac{g}{V} = \frac{P \times MM}{RT}$$

$$MM = \frac{g \times R \times T}{P \times V}$$

DALTON'S LAW OF PARTIAL PRESSURES

$$P_{Total} = P_{component_1} + P_{component_2}$$

$$P_{Total} = x_{component_1} P°_{component_1} + x_{component_2} P°_{component_2}$$

Chapter 5: Reaction Kinetics

RATE EQUATIONS

$$2A + 4B \Rightarrow 5C + 3D$$

$$\text{Rate} = -\frac{1}{2}\frac{\Delta[A]}{\Delta t} = -\frac{1}{4}\frac{\Delta[B]}{\Delta t} = +\frac{1}{5}\frac{\Delta[C]}{\Delta t} = +\frac{1}{3}\frac{\Delta[D]}{\Delta t}$$

$$3A + B \Rightarrow 2C + D$$

$$\text{Rate} = -\frac{1}{3}\frac{\Delta[A]}{\Delta t} = -\frac{1}{1}\frac{\Delta[B]}{\Delta t} = +\frac{1}{2}\frac{\Delta[C]}{\Delta t} = k[A]^m[B]^n$$

ARRHENIUS EQUATION

$$k = Ae^{-\frac{E_a}{RT}}$$

where: A is a constant
E_a is the activation energy.

$$\ln\frac{k_2}{k_1} = -\frac{E_a}{R}\left[\frac{1}{T_2} - \frac{1}{T_1}\right]$$

Chapter 6: Equilibrium

EQUATIONS RELATING K_C AND K_P

$$K_c = K_p \left[\frac{1}{RT}\right]^{\Delta n = ((\text{moles of product}) - (\text{moles of reactant}))}$$

$$K_p = K_c [RT]^{\Delta n = ((\text{moles of product}) - (\text{moles of reactant}))}$$

VAN'T HOFF EQUATION

$$\ln \frac{K_2}{K_1} = -\frac{\Delta H°}{R}\left[\frac{1}{T_2} - \frac{1}{T_1}\right]$$

CLAUSIUS CLAPEYRON EQUATION

$$\ln \frac{P_2°}{P_1°} = -\frac{\Delta H°_{vap}}{R}\left[\frac{1}{T_2} - \frac{1}{T_1}\right]$$

IONIZATION CONSTANTS

$K_a \times K_b = K_w = 1.0 \times 10^{-14}$

BUFFERS

$$[H^+] = K_a \times \frac{[HA]}{[A^-]}$$

$$pH = pK_a + \log \frac{[A^-]}{[HA]} \qquad \text{(Henderson-Hasselbach equation)}$$

NEUTRALIZATION

$M_{base} V_{base} = M_{acid} V_{acid}$

Chapter 7: Thermodynamics

INTERNAL ENERGY CHANGE, HEAT AND WORK

$\Delta E = q + w \qquad q = 4.184 \times m_{H_2O} \times \Delta t_{H_2O} \qquad w = P\Delta V = \Delta n_{gas} RT$

ENTHALPY CHANGE

$\Delta H = \Delta E + w = \Delta E + (\Delta n)RT \qquad \Delta H° = \sum \Delta H_f° \text{ (Products)} - \sum \Delta H_f° \text{ (Reactants)}$

ENTROPY AND ENTROPY CHANGE

$\Delta S° = \int c_p \, dT = q_p / T \qquad \Delta S° = \sum S°\text{products} - \sum S°\text{reactants}$

FREE ENERGY CHANGE

$\Delta G° = \Delta H° - \dfrac{T\Delta S°}{1000} \qquad \Delta G° = \sum \Delta G_f° \text{ (Products)} - \sum \Delta G_f° \text{ (Reactants)}$

$\Delta G° = -RT \ln K \qquad \Delta G° = -nF\mathcal{E}°$

Chapter 8: Electrochemistry

NERNST EQUATION

At 25°C: $\qquad \mathcal{E} = \mathcal{E}° - \dfrac{0.0592}{n} \log Q$

Data that may be useful in solving the problems.

B. Standard Reduction Potentials

Standard half-cell potentials at standard conditions. (1.0 M solutions 1.0 atmosphere pressure and 25°C)		
$Li^+ + e^-$	$\Rightarrow$ Li	–3.04 volts
$K^+ + e^-$	$\Rightarrow$ K	–2.92 volts
$Na^+ + e^-$	$\Rightarrow$ Na	–2.71 volts
$Mg^{2+} + 2e^-$	$\Rightarrow$ Mg	–2.37 volts
$Al^{3+} + 3e^-$	$\Rightarrow$ Al	–1.66 volts
$2H_2O + 2e^-$	$\Rightarrow$ $H_2 + 2OH^-$	–0.83 volts
$Zn^{2+} + 2e^-$	$\Rightarrow$ Zn	–0.76 volts
$Fe^{2+} + 2e^-$	$\Rightarrow$ Fe	–0.45 volts
$Cd^{2+} + 2e^-$	$\Rightarrow$ Cd	–0.40 volts
$Co^{2+} + 2e^-$	$\Rightarrow$ Co	–0.28 volts
$Ni^{2+} + 2e^-$	$\Rightarrow$ Ni	–0.26 volts
$Sn^{2+} + 2e^-$	$\Rightarrow$ Sn	–0.14 volts
$Pb^{2+} + 2e^-$	$\Rightarrow$ Pb	–0.13 volts
$2H^+ + 2e^-$	$\Rightarrow$ H_2	0.00 volts
$Cu^{2+} + 2e^-$	$\Rightarrow$ Cu	0.34 volts
$I_2 + e^-$	$\Rightarrow$ $2I^-$	0.54 volts
$Hg^{2+} + 2e^-$	$\Rightarrow$ Hg	0.91 volts
$Ag^+ + e^-$	$\Rightarrow$ Ag	0.80 volts
$Br_2 + 2e^-$	$\Rightarrow$ $2Br^-$	1.09 volts
$O_2 + 4H^+ + 4e^-$	$\Rightarrow$ $2H_2O$	1.23 volts
$Cl_2 + 2e^-$	$\Rightarrow$ $2Cl^-$	1.36 volts
$Au^{3+} + 3e^-$	$\Rightarrow$ Au	1.50 volts
$F_2 + 2e^-$	$\Rightarrow$ $2F^-$	2.87 volts

C. Vapor Pressure of Water

Temperature, °C	Vapor Pressure of Water mmHg	Temperature, °C	Vapor Pressure of Water mmHg
0	4.6	28	28.1
10	9.2	29	29.8
15	12.7	30	31.5
20	17.4	35	41.8
21	18.5	40	55.0
22	19.7	50	92.2
23	20.9	60	149.2
24	22.2	70	233.7
25	23.6	80	355.1
26	25.1	90	525.8
27	26.5	100	760.0

D. Constants

Universal Gas Constant R	8.314 J mol^{-1} K^{-1}
	0.0821 L atm mol^{-1} K^{-1}
	62.4 L mmHg mol^{-1} K^{-1}
	8.314 V C mol^{-1} K^{-1}
1 Faraday	96,500 coulombs
1 calorie	4.184 joules
1 electron volt atom^{-1}	96,500 kilojoules mol^{-1}
Speed of light (vacuum)	2.998 x 10^8 m s^{-1}
ln$_e$ x	2.303 log$_{10}$ x
Planck's constant (h)	1.38 x 10^{-23} Joule K^{-1}
Avogadro's number	6.022 x 10^{23} molecules mole^{-1}
At 25°C	$\frac{RT}{nF}$ ln Q or $\frac{0.0591}{n}$ ln Q

Index

acid-base anhydrides	178
acid-base equilibrium	162, 165
acid-base titration	162, 166
activated complex	50
activation energy	49, 52, 128, 129, 130, 194
alkali metals	5
alpha particle	8, 105
ampere	91, 150, 168
amphiprotic	68
angular momentum	103
anion	9, 25, 63, 94
anode	90, 94, 150, 151, 168
anti-bonding electrons	110
anti-bonding orbital	21
aqua regia	105
Arrhenius equation	51
atom	1
atomic number	1
atomic orbitals	2
atomic radius	7
atomic weight	2
atomic weights	27
attractive force	18, 36, 108, 195
aufbau	3
balancing	
in acid solution	88
in basic solution	89
redox reactions	87
beta particle	8, 105
boiling temperature	121
boiling point elevation	34
bomb calorimeter	77, 141
bond angle	191
bond distance	110
bond energy	81, 110, 143
bond order	21
bonding electrons	110
bounce back	64, 65
brittleness	9
Brönsted-Lowry theory	68
buffer	71, 136
calorimetry	76
catalyst	49, 58, 128, 129
cathode	90, 94, 150, 151, 168
cation	9, 25, 63, 104
cell potential or voltage	86, 168, 192
chain-ending reactant	128
chain-maintaining agent	128
changes of state	38
chemical bonding	9, 108
chemical or voltaic cell	90, 94, 151, 168
Clausius Clapeyron equation	59
colligative properties	34
collisions	44, 49, 127, 130
combination reaction	178, 183
common ion effect	62
complete equation	177
complex ions	24, 192
concentration	58
condensation	121
conductivity	9
constants	207
coordinate covalent bond	24, 108
coordination chemistry	24
coordination compound	110, 180
covalent bond	9, 108
critical point	39
critical pressure	39
critical temperature	39
crystal lattice	36
crystalline	9, 108
crystallization	121
Dalton's law of partial pressures	42, 43, 55
decomposition	179, 183
density	115, 142
deuteron	8
diamagnetism	21, 191
diatomic elements	175
dipole	19, 120
dipole attractions	36
dispersion forces	36
displacement	178
dissociation constant	68
ductility	9
effusion	39, 123
electrical conductivity	9, 108

electrochemical cells	90
electrochemistry	168
electrode	90, 94, 150, 168
electrode, inert	90
electrolysis	90, 150, 168
electrolyte	90, 94
electron	
affinity	6, 7, 105, 108, 191
hybridization	24
sea	9
stability	10
transfer	87
transitions	103
unbonded	11
electron-dot diagram	12
electronegativity	9, 108
electrostatic attraction	9
electrostatic forces	120
empirical formula	28, 113, 114, 169
endothermic reaction	50, 104, 144
energy diagram	128, 129
energy of activation	169
energy states	2
enthalpy	76, 83, 141
entropy	76, 82, 83, 144, 170
equilibrium	
concentrations	162
constant	55, 57–59, 76, 133, 161, 168
K_a	68, 135, 162
K_b	68, 162
K_c	55, 133
K_p	55, 133, 161
K_{sp}	61, 161
K_w	68
law	55
vapor pressure	85
equivalence point	72, 136, 162
exothermic reaction	50, 104
Faraday	91, 150
first order reactions	48
free energy	76, 83, 144, 170
free energy and equilibrium	84
freezing point constant	34, 169
gas density	41
gaseous equilibrium	161, 163
geometric shape	109
Graham's law	39
groups	5
half-life	8, 48, 105
half-reaction	87, 149
heat	76
capacity	141
of combustion	80, 143
of formation	80, 143, 170
of fusion	144
of reaction	76, 80, 81, 129, 141, 143, 170
of vaporization	60
Hess' law	78, 80
Henderson-Hasselbach equation	71
Hund's rule	3, 21, 191
hybrids	23, 110
hydration energy	104, 120
hydride ion	12
hydrogen bond	18, 36, 108, 120
hydrolysis	179
intermediate	53
internal energy	76, 141
ideal gas law	40
ion structure	5
ionic	36, 120
ionic bond	9, 108
ionic solids	8
ionization constant	68
ionization energy	6, 7, 102, 104
ions	2
isomers	26
isotope	1, 103, 105
joules	76
kernel	9
kinetic energy	120, 130
kinetic energy distribution	129
kinetic molecular theory	195
kinetic molecular theory of gases	39

LeChatelier's principle	58, 70, 98	octet formulas	10
Lewis electron-dot structures	10–12, 108, 192	octet, incomplete	14
ligand	24, 25, 108, 180, 110, 192	odd-electron molecules	14
limiting reactants	31	orbital	3, 103, 104
line structure	11	overall order	47
linear	15, 16	oxidation	90
London dispersion forces	19, 36	oxidation number	87
lone-pair repulsion	15	oxidation states	104
lone-pair electrons	10	oxidation-reduction	87
		oxidized	181
malleability	9	oxidizer	181, 182
mass number	1	oxidizing agent	87, 96, 149, 181, 192
Maxwell-Boltzmann distribution	49		
mechanism	128, 130, 169	paramagnetic	21, 104, 110, 191
melting temperature	39, 108	partial pressure	42, 55, 127, 133, 161
metallic	36	Pauli exclusion principle	3, 21, 191
metallic bond	9, 104, 108	percent dissociation	57
metallic character	7	periodic trends	7
metallic solids	18	periods	5
molality	32, 33, 115	pH	67, 68, 135, 136, 162
molar mass	27, 41, 113, 122, 169	phase change diagram	38
molar solubility	62, 63, 161	pi bonding	23, 110, 192
molarity	32, 33, 55, 115, 133, 170	polar	120
mole		polar solvents	9
concept	27	polyatomic anions and cations	175
fraction	32, 33, 115	positive ions	5
link	27, 28, 30, 56	potential energy	120
molecular formula	28, 114, 169	potential energy diagram	50
molecular orbital theory	21, 110	precipitation reactions	177
molecules	18	pressure	58
monoprotic	72	proton	1, 8
Nernst equation	98	quantum number	2, 103
net driving force	83	magnetic	3
net ionic equation	177, 193	orbital	2
network covalent	18, 36, 120	principal	2
neutralization	72	spin	3
neutron	1, 8		
noble metal	105	radius of atoms and ions	5, 104
nondipoles	19	radioactive	102
nonspontaneous	83	radioactive decay	8
nuclear charge	105	Raoult's law	34, 35, 42
nuclear chemistry	8		
octahedral	15, 17, 108, 109		

rate
- constant   127
- determining step   127
- equation   44, 45, 128
- law constant   194
- law equation   126, 130, 169, 194
- of reaction   130, 169

reaction
- mechanism   53
- reaction order   44
- reaction rate   126
- reaction stoichiometry   30

real gases   195
redox   87, 181
reduced   181
reducer   181, 182
reducing agent   87, 96, 150, 181, 192
reduction   90
reduction potential   168
replacement   182
resonance   13, 109

salt bridge   94
saturated solution   64, 161
selective precipitation   63
shape   191
shell   2
sigma bonding   23–24
sigma bonds   23–24, 110, 192
slow step   53
solubility equilibrium   61, 161, 164
solubility product constant (K_{sp})   61, 161
solubility rules for salts   176
solute   32
solution   32
solution stoichiometry   34
solvent   32
specific heat   142
specific rate law constant   44, 169
spontaneous reaction   76, 83, 94, 144
square planar   109
standard heats of combustion   81
standard reduction potential   152
standard reduction potentials   96, 206
stoichiometry   27

strong acids   66
strong bases   66
structural formulas   11
sublimation   36, 38, 121
subshell   2, 3

tetrahedral   15, 16, 109, 191
thermal   108
thermochemical reaction   142
titrating   136
titration   72, 162, 170
transition element   24, 104, 108
trigonal bipyramidal   15, 17, 109
trigonal planar   15, 16
trigonal pyramidal   109
triple point   39
triton   8

unsaturated   64

valence   10, 102, 108
valence electrons   11
valence shell   9, 108
VSEPR theory   15, 21
van der Waal's force   18, 36, 120
van't Hoff equation   59
vapor pressure   34, 36, 38, 60, 120, 123
vaporization   121
velocity of molecules   130
voltage   86

weak acids   68
weak bases   68
work   76, 77

Periodic Table of the Elements

1A	2A											3A	4A	5A	6A	7A	
1 H 1.008																	2 He 4.003
3 Li 6.941	4 Be 9.012											5 B 10.81	6 C 12.01	7 N 14.01	8 O 16.00	9 F 19.00	10 Ne 20.18
11 Na 22.99	12 Mg 24.31											13 Al 26.98	14 Si 28.09	15 P 30.97	16 S 32.06	17 Cl 35.45	18 Ar 39.95
19 K 39.10	20 Ca 40.08	21 Sc 44.96	22 Ti 47.90	23 V 50.94	24 Cr 52.00	25 Mn 54.94	26 Fe 55.85	27 Co 58.93	28 Ni 58.70	29 Cu 63.55	30 Zn 65.38	31 Ga 69.72	32 Ge 72.59	33 As 74.92	34 Se 78.96	35 Br 79.90	36 Kr 83.80
37 Rb 85.47	38 Sr 87.62	39 Y 88.91	40 Zr 91.22	41 Nb 92.91	42 Mo 95.94	43 Tc (98)	44 Ru 101.1	45 Rh 102.9	46 Pd 106.4	47 Ag 107.9	48 Cd 112.4	49 In 114.8	50 Sn 118.7	51 Sb 121.8	52 Te 127.6	53 I 126.9	54 Xe 131.3
55 Cs 132.9	56 Ba 137.3	57 La* 138.9	71 Hf 178.5	73 Ta 180.9	74 W 183.9	75 Re 186.2	76 Os 190.2	77 Ir 192.2	78 Pt 195.1	79 Au 197.0	80 Hg 200.6	81 Tl 204.4	82 Pb 207.2	83 Bi 209.0	84 Po (209)	85 At (210)	86 Rn (222)
87 Fr (223)	88 Ra 226.0	89 Ac** (227)	104 Unq (261)	105 Unp (262)	106 Unh (263)	107 Uns (262)	108 Uno (265)	109 Une (267)									

*Lanthanides (Rare Earths)

58 Ce 140.1	59 Pr 140.9	60 Nd 144.2	61 Pm (145)	62 Sm 150.4	63 Eu 152.0	64 Gd 157.3	65 Tb 158.9	66 Dy 162.5	67 Ho 164.9	68 Er 167.3	69 Tm 168.9	70 Yb 173.0	71 Lu 175.0

**Actinides (Transuranium)

90 Th 232.0	91 Pa (231)	92 U 238.0	93 Np (237)	94 Pu (244)	95 Am (243)	96 Cm (247)	97 Bk (247)	98 Cf (251)	99 Es (252)	100 Fm (257)	101 Md (258)	102 No (259)	103 Lr (260)